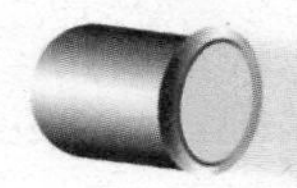

Spotlight on Georgia Performance Standards

HSP Georgia Science

Interactive Text

Harcourt
SCHOOL PUBLISHERS

Visit *The Learning Site!*
www.harcourtschool.com

Copyright © by Harcourt, Inc.

All rights reserved. No part of this publication may be reproduced or transmitted in any form or by any means, electronic or mechanical, including photocopy, recording, or any information storage and retrieval system, without permission in writing from the publisher.

Requests for permission to make copies of any part of the work should be addressed to School Permissions and Copyrights, Harcourt, Inc., 6277 Sea Harbor Drive, Orlando, Florida 32887-6777. Fax: 407-345-2418.

HARCOURT and the Harcourt Logo are trademarks of Harcourt, Inc., registered in the United States of America and/or other jurisdictions.

Printed in the United States of America

ISBN 13: 978-0-15-378395-1
ISBN 10: 0-15-378395-8

17 18 19 20 1421 16 15 14 13 4500455058

If you have received these materials as examination copies free of charge, Harcourt School Publishers retains title to the materials and they may not be resold. Resale of examination copies is strictly prohibited and is illegal.

Possession of this publication in print format does not entitle users to convert this publication, or any portion of it, into electronic format.

Unit A Earth Science

Chapter 2 Planets and Stars

Chapter 3 Weather

Unit B Physical Science

Chapter 4 Sound and Light

Chapter 5 Forces and Motion

Chapter 6 Simple Machines

Unit C Life Science

Chapter 7 Food Energy in Ecosystems

Chapter 8 Adaptations for Survival

Planets and Stars

The Big Idea We can predict where the sun, the moon, the stars, and the planets will be in the sky.

On this page, record what you learn as you read the chapter.

Essential Question

How do earth and its moon move?

____

____

____

Essential Question

What can we see in the sky?

____

____

____

Essential Question

What objects are in the solar system?

____

____

____

Quick and Easy Project

Tell Time by Using the Sun

Materials:

- small ball of clay
- pencil
- ruler
- 15-cm x 20-cm cardboard

Procedure

1. On the cardboard, draw a half circle with a 7-cm radius. Put the clay in the center of the straight side.
2. Stand the pencil in the clay with the point up. You have made a sundial.
3. Put your sundial on a sunny windowsill, with the straight side along the window. Every hour, mark the pencil's shadow on the cardboard. Write the time at the mark.
4. On the next sunny day, use your sundial to tell time.

Draw Conclusions

1. How does your sundial work?
2. Would it work in any sunny window?

   Why or why not?

Independent Inquiry

Make a Spiral Galaxy

A spiral galaxy gets its name from the shape formed by many orbiting stars. Use a paper punch to punch out many "stars" from construction paper. Fill a large bowl halfway with water. Pour the stars into the water. How can you make them form a "spiral galaxy"?

Georgia Performance Standards

S4E2a Explain the day/night cycle of Earth using a model.

S4E2b Explain the sequence of the phases of the moon.

S4E2c Demonstrate the revolution of the earth around the sun and the earth's tilt to explain the seasonal changes.

Vocabulary Activity

The way Earth and the moon move affects many of the patterns we observe in the sky. Use the pictures on pages 2 and 3 to draw one picture that shows the orbits of Earth and its moon. Use the vocabulary terms to label your drawing.

Orbits of Earth and Its moon

Student eBook
www.hspscience.com

Lesson 1

How Do Earth and Its Moon Move?

VOCABULARY
rotate
axis
revolve
orbit
moon
phase

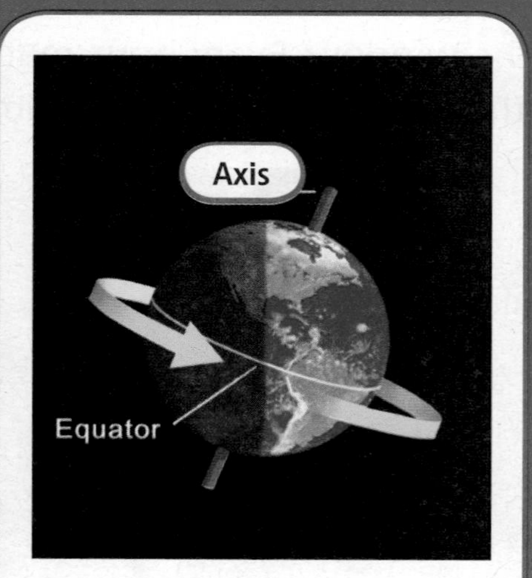

Earth **rotates,** or turns, on its **axis**, an imaginary line that runs through both poles.

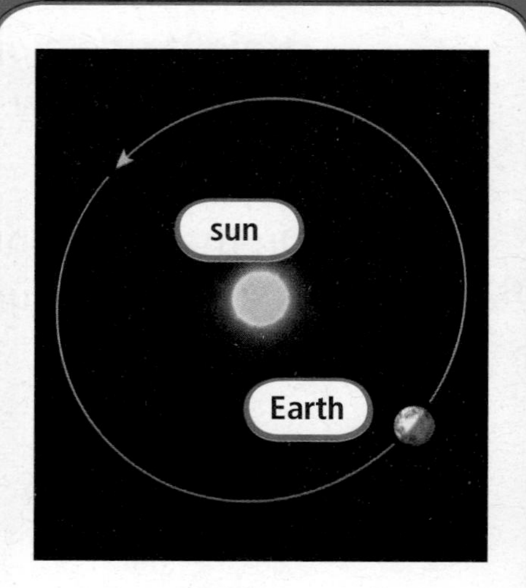

Earth's **orbit** is the path Earth travels as it **revolves** around the sun.

The **moon** is a large object that travels around Earth.

The moon's **phases** are the different shapes the moon appears to have.

Insta-Lab

Sunrise, Sunset

1. Model how the sun and Earth interact to cause day and night.
2. Use a flashlight and a ball. Shine the flashlight on one side of the ball.
3. Where on the ball are sunrise and sunset represented?

____

____

4. How does this model the day/night cycle of Earth?

____

____

____

When you sequence things, you put them in order. How can you sequence day and night?

____

____

1. Explain what causes day and night.

____

____

2. How often does the sequence of day and night repeat?

____

____

Day and Night

Every day, the sun seems to rise in the east. It reaches its highest point around noon. Later, it appears to set in the west. Darkness follows. We have this cycle of day and night because Earth **rotates**, or spins, on its axis. Earth's **axis** is an imaginary line that passes through the North and South Poles. When a place on Earth faces the sun, that place has day. When that place faces away from the sun, it has night. Earth's cycle of daylight and darkness repeats about every 24 hours.

The sun doesn't really rise, move across the sky, and then set. It only seems to because of Earth's rotation.

Much of the United States is within one of four time zones. Time zone lines aren't perfectly straight, partly because of state boundaries.

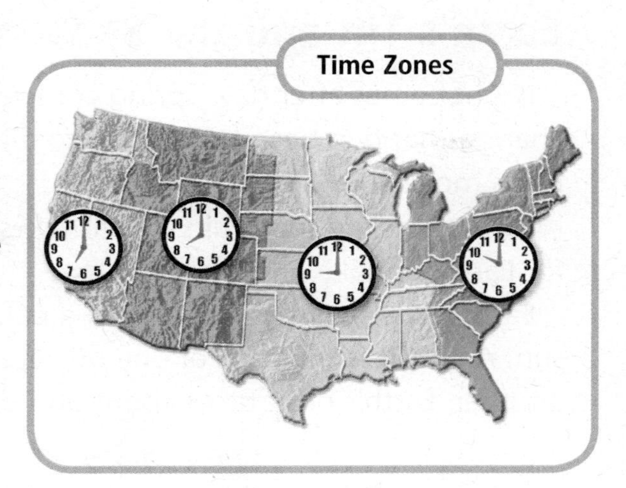

Our system of time is based on this 24 hour cycle. Long ago, people in different places used local time. Local time was based on when sunrise and sunset happened in that place. Because sunrise and sunset occur at different times in different places, there was no exact way to tell time.

In 1884, people set up 24 times zones around the world. Each time zone represents one of the hours in the day. All the places within a time zone have the same time. The United States has seven time zones, from Puerto Rico in the east to Hawaii in the west.

Suppose it is 6 PM in Georgia. You're just about to have dinner. You call a friend in Oregon. It's only 3 PM there. Your friend may just be getting home from school.

✓ Quick Check

1. Look at the map on this page. Circle the time zone where you live.
2. How many time zones are there in the United States?

   ____

3. Suppose you fly from Georgia to California. Tell about the sequence of time zones you travel through.

   ____

   ____

   ____

   ____

Quick Check

When you sequence things, you put them in order. How can you sequence the seasons?

____

____

____

1. What is the word for the path of one object in space around another?

____

2. How long does it take Earth to revolve around the sun one time?

____

____

3. In what two ways does Earth move?

____

____

Earth's Tilt and the Seasons

Night comes after day. Spring comes after winter. These changes happen because of the two different ways that Earth moves.

Earth turns around on its axis. It takes about 24 hours for Earth to make one full turn.

Earth also **revolves**, or travels in a path around the sun. An **orbit** is the path of one object in space around another. Earth's orbit takes about 365 days, or one year.

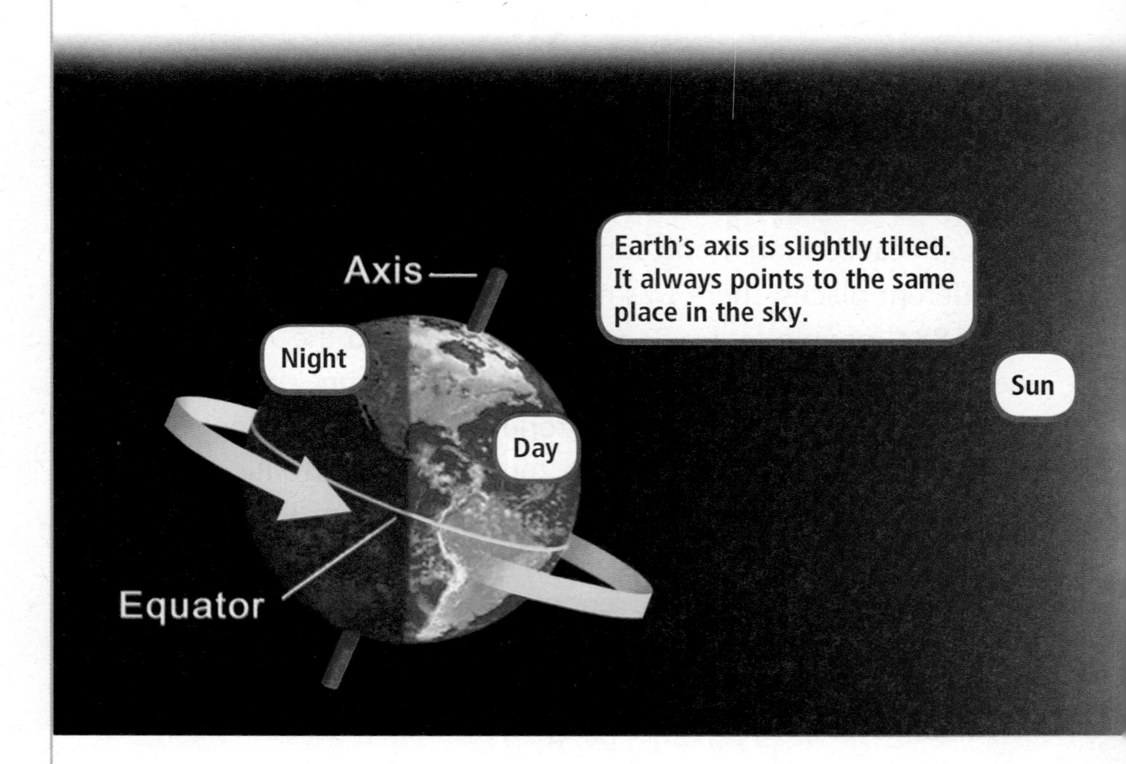

As Earth revolves, one part is tilted toward the sun. This part receives more heat and light. It is summer there. The other part of Earth is tilted away from the sun. That part of Earth receives less heat and light. It is winter there.

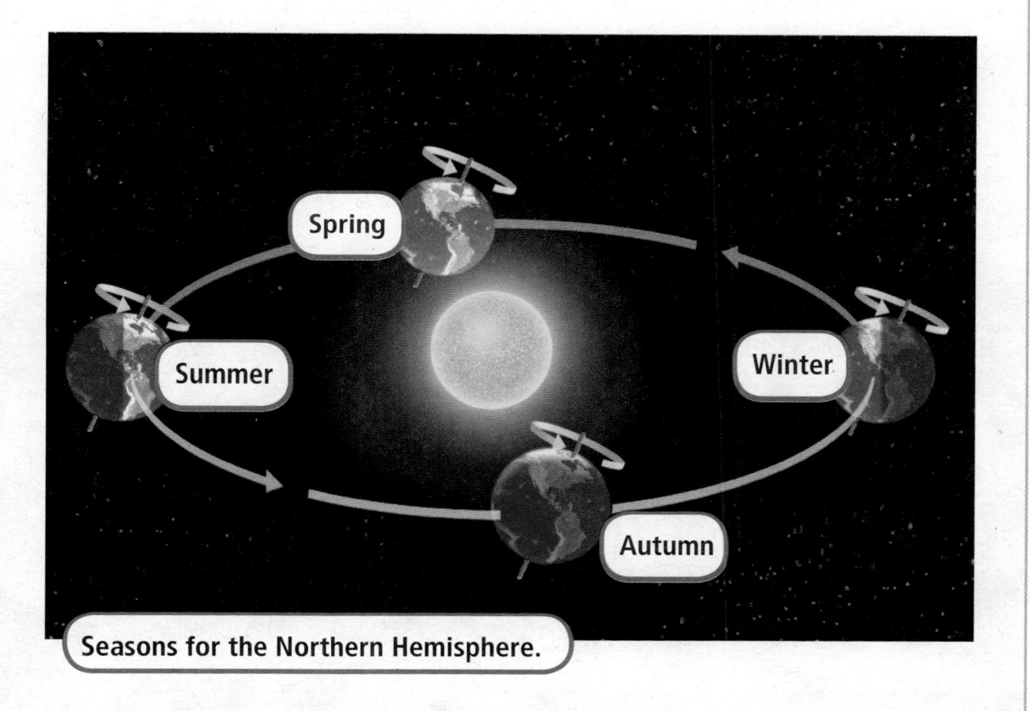

Seasons for the Northern Hemisphere.

The seasons change as Earth orbits the sun. This happens because the part of Earth that is tilted toward the sun changes.

Quick Check

1. Which part of Earth receives more heat and light?

____

____

2. Place an X on the season in the image where the Northern Hemisphere is receiving the least heat and light.

3. The Northern Hemisphere is tilted away from the sun. What season is it in Georgia?

____

Tell what happens next.

____

____

✓ Quick Check

Focus Skill When you sequence things, you put them in order. How can you sequence the moon's phases?

____

____

1. Fill in the chart. Check the column to tell whether each statement applies to that moon phase. Some rows may have more than one check.

	New Moon	First Quarter Moon	Full Moon	Third Quarter Moon
Moon appears half lit				
Moon appears unlit				
Moon appears completely lit				
Farthest from the sun				

Moon Phases

The **moon** is a large object that revolves around Earth. It takes about $29\frac{1}{2}$ days for the moon to complete one orbit. The moon's light is reflected sunlight.

Over $29\frac{1}{2}$ days, the moon's shape seems to change. These different shapes are called **phases**. These phases follow a regular pattern, or cycle. What causes these changes? As the moon orbits Earth, we see different amounts of its lit surface.

The moon's phases as seen from Earth

When we see all of the moon's lit side, this is a full moon. After that you see less of the moon each day. After about 15 days, you cannot see any of the lit side. This is a new moon. For about 15 days after the new moon, we see more and more of the moon. Finally, we see all of the lit side again. Then the cycle repeats.

Lesson Review

Complete these sequence statements.

1. During one part of the moon's cycle, you see all of the lit side. This is called a ___________.
2. The moon's shape seems to change in a pattern because as the moon orbits Earth we see different amounts of its ___________ surface.
3. Label the seasons for the Northern Hemisphere. Draw arrows to show sequence.

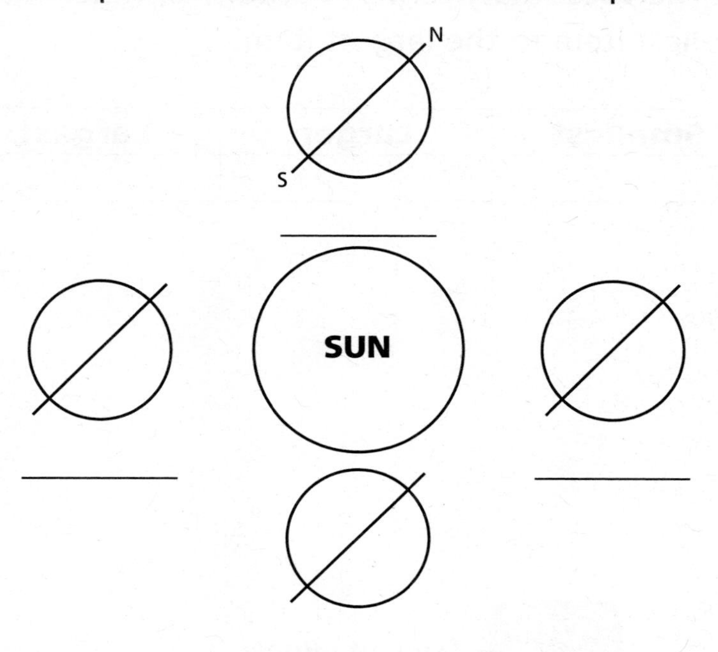

Georgia Performance Standards

S4E2d Demonstrate the relative size and order from the sun of the planets in the solar system.

S4E1d Identify how technology is used to observe distant objects in the sky.

Vocabulary Activity

In this lesson, you'll learn about different objects in space.

List the vocabulary terms in order from the smallest item to the largest item.

Smallest	Larger	Largest

Go online Student eBook www.hspscience.com

Lesson 2

VOCABULARY
solar system
planet
comet

What Objects Are in the Solar System?

A **solar system** is a group of objects in space that travel around a star. The sun is the star in the center of our solar system.

A **planet** is a large object in space that moves around a star. Earth is one of eight planets in our solar system.

A **comet** is a ball of rock, ice, and frozen gas.

Insta-Lab

Planet Sizes

1. Line up a large marble, a table tennis ball, and a basketball in a row.

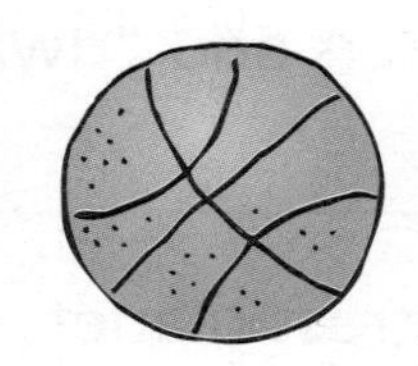

2. Observe the size of each item in comparison to the other two items.
3. How did the sizes of the materials compare?

____

____

____

4. When comparing sizes, which planet could each item represent?

____

____

____

Quick Check

You compare by telling how things are alike. You contrast ways things are different. Compare planets and "dwarf planets."

___

Contrast planets and "dwarf planets."

___

___

1. How are inner planets and outer planets different?

___

___

2. Tell how planets and moons are alike and different.

___

___

___

Our Solar System

A **solar system** is a group of objects in space that travel around a star. The sun is the star in the center of our solar system.

Our solar system has different kinds of objects. There are planets, "dwarf planets," moons, and asteroids. A **planet** is a large object that revolves around a star in a clear orbit. A "dwarf planet" also revolves around a star, but its path is not clear of other objects. A moon is smaller. It revolves around a planet.

Our solar system has eight planets. They all orbit the sun. Scientists put these planets into two groups. The inner planets are closer to the sun. The outer planets are farther from the sun. These two groups are separated by a ring of small, rocky objects. These objects are asteroids.

✓ Quick Check

1. Which planet is the most like Earth in size?

____

2. How are Saturn and Uranus alike?

____

3. Contrast Neptune and Mars.

____

____

____

4. Number the planets by their order from the sun. The first one is done for you.

_____ Earth	_____ Jupiter	_____ Venus
_____ Mars	_____ Saturn	__1__ Mercury
_____ Neptune	_____ Uranus	

✓ Quick Check

Focus Skill You compare by telling ways things are alike. You contrast by telling ways things are different. Compare the inner planets.

____

____

Contrast Mercury and Mars.

____

1. Tell how the inner planets differ.

____

____

____

2. How is Earth different from the other inner planets?

____

____

The Inner Planets

The four inner planets are Mercury, Venus, Earth, and Mars. These planets are alike in many ways. Each has a rocky surface. All are smaller than the outer planets. None has more than two moons.

Mercury gets very hot. Mars gets very cold. Only Earth has water on its surface. It is also the only planet that has a lot of oxygen. Water and oxygen let plants and animals live on Earth.

The Outer Planets

The four outer planets are far from the sun. They are Jupiter, Saturn, Uranus, and Neptune.

Jupiter, Saturn, Uranus, and Neptune are alike in many ways. They are large. They are made mostly of gases. These planets are often called the gas giants. These planets have many moons.

Many spacecraft have been sent to study the outer planets. These include *Voyager 1* and *2*, *Galileo*, and *Cassini*.

✓ Quick Check

1. Compare the outer planets.

____

____

____

2. How have scientists used technology to observe the outer planets?

____

____

3. How are the inner planets different from the outer planets?

____

____

____

____

Quick Check

You compare by telling ways things are alike. Compare the "dwarf planets" and the inner planets.

____

____

You contrast by telling ways things are different. Contrast "dwarf planets" and the inner planets.

____

____

____

1. Tell whether each statement about "dwarf planets" is true or false.

   ______ They are larger than the outer planets.

   ______ There may be hundreds of them.

   ______ They are smaller than the inner planets.

   ______ They are made of gases.

What About Pluto?

For a long time, Pluto was counted as an outer planet. But some scientists weren't so sure it should be. Pluto is very different from Jupiter, Saturn, Uranus, and Neptune. Pluto is made of rock and ice. The outer planets are made of gases. The outer planets are large, but Pluto is small. The inner and outer planets revolve around the sun in clear paths. This means that their orbits are clear of other large objects. This isn't true of Pluto.

Pluto is now called a "dwarf planet." Two other "dwarf planets" are Ceres and Eris. From largest to smallest, they are Eris, Pluto, and Ceres. Scientists say that there may be hundreds of "dwarf planets" in our solar system.

Our solar system has eight planets and three named "dwarf planets."

Small Solar System Bodies

Two other kinds of objects revolve around the sun. Asteroids are made of rock and metal. Most asteroids orbit between Mars and Jupiter.

A **comet** is a ball of rock, ice, and frozen gas. Most are smaller than asteroids. Comets can change when they come near the sun. Some of the frozen matter turns to gas. This gas looks like a tail.

A comet's fiery tail

Lesson Review

Complete these compare-and-contrast statements.

1. The largest planet in our solar system is __________.

2. __________ is the largest known "dwarf planet."

3. Fill in this graphic organizer.

Inner Planets	Outer Planets
What's Different? Closer to _____ _____ Small with rocky surfaces No more than _____ _____	**What's Different?** ________ from the sun Have many ________ Rings of dust, ice, and rock.

What's the same?
Part of the ________ ________
Planets orbit ________ ________

Georgia Performance Standards

S4E1a Recognize the physical attributes of stars in the night sky such as number, size, color, and patterns.

S4E1b Compare the similarities and differences of planets to the stars in appearance, position, and number in the night sky.

S4E1c Explain why the pattern of stars in a constellation stays the same, but a planet can be seen in different locations at different times.

S4E1d Identify how technology is used to observe distant objects in the sky.

Vocabulary Activity

Use the vocabulary words to fill in the blanks.

1. A ________________ is a group of stars that forms an imaginary picture in the sky.
2. The ______________ is everything in space.
3. A huge system of gases, dust, and stars is called a ___________.
4. A ___________ is a huge ball of very hot gas.

Student eBook
www.hspscience.com

Lesson 3

What Can We See in the Sky?

VOCABULARY
star
sun
constellation
galaxy
universe

A **star** is a huge ball of very hot gas.

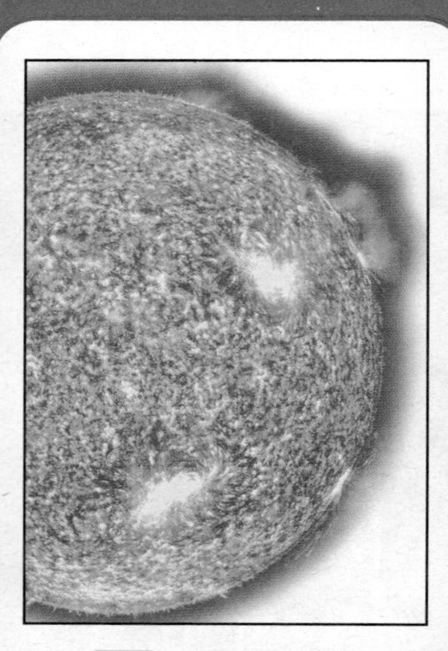

The **sun** is the star in the center of our solar system.

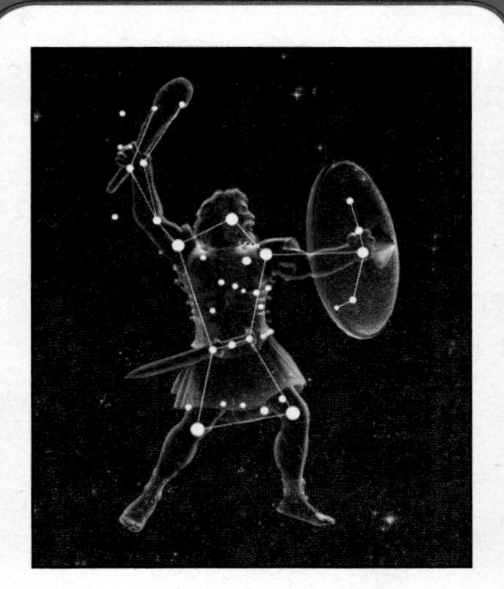

A **constellation** is a group of stars that form an imaginary picture in the sky.

A **galaxy** is a huge system of gases, dust, and stars.

The **universe** is everything in space.

Insta-Lab

Model a Constellation

1. Use a pencil to poke a "constellation" in the center of a piece of aluminum foil.
2. Use a rubber band to fasten the foil to the end of a paper towel tube.
3. Look through the tube. What do the holes in the foil represent as you look through the tube?

____

____

4. Turn the tube while you look through it. What motion are you modeling as you turn the tube?

____

✓ Quick Check

Focus Skill The main idea on these two pages is There are different ways to group stars. Details tell more about the main idea. Underline two details about how stars can be grouped.

1. From page 21, which type of star group has the smallest number of stars?

____

____

2. Fill in the chart. Tell the temperatures of each type of star.

Temperature	Star Color
Hottest	
Cooler	
Coolest	

3. What color star is the sun?

____

4. What are two ways stars can be different?

____

The Sun and Other Stars

A **star** is a huge ball of very hot gas. Stars may be different sizes and colors. Red stars are the coolest. Blue stars are the hottest. Yellow stars are between the hottest and the coolest.

The **sun** is the star at the center of our solar system. It is a medium-size, yellow star. Most energy on Earth comes from the sun.

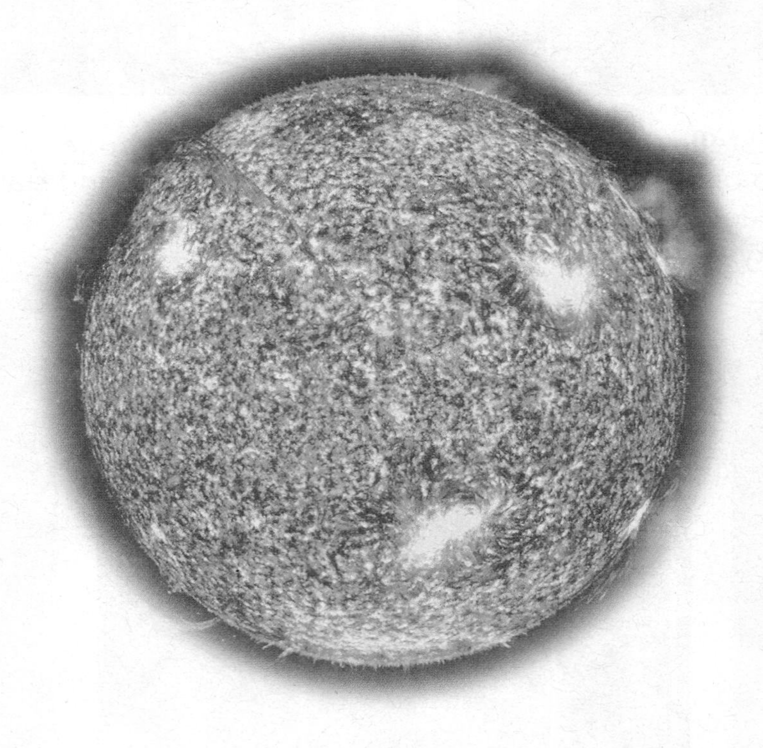

The sun is more than 1 million kilometers (621,000 miles) in diameter.

Groups of Stars

The Big Dipper is part of a constellation. A **constellation** is a group of stars that make an imaginary picture.

A **galaxy** has billions of stars. A galaxy is a huge system of gases, dust, and stars. Our solar system is on the edge of the Milky Way galaxy. Scientists use telescopes to observe different galaxies.

The **universe** is everything in space. It has billions of galaxies.

What are two ways that people classify groups of stars?

▼ **Galaxy**

▲ **This constellation is called Orion.**

✓ Quick Check

1. What is a detail about this main idea: "A galaxy is a huge system of gases, dust, and stars."?

____

____

2. What are two ways people classify groups of stars?

____

____

3. In the space below, draw and name a constellation that you designed.

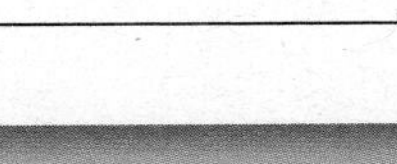

Quick Check

Focus Skill The main idea on these two pages is Objects seem to move across the sky. Details tell more about the main idea. Underline two details about how objects seem to move across the sky.

1. Name two objects that appear to move across the sky.

____

____

2. How can you tell a star from a planet in the night sky?

____

____

3. Why does the pattern of stars in a constellation stay the same, while planets can be seen in different locations?

____

____

Seasonal Star Positions

Each day the sun appears to move in the sky. But the sun does not move. It only seems to move as Earth rotates on its axis.

At night the stars seem to move. Their positions appear to change from season to season. But this is due to Earth's movement. As Earth revolves around the sun, we see different parts of space at different times of year.

▲ **Constellations in the summer sky**

▲ **Constellations in the winter sky**

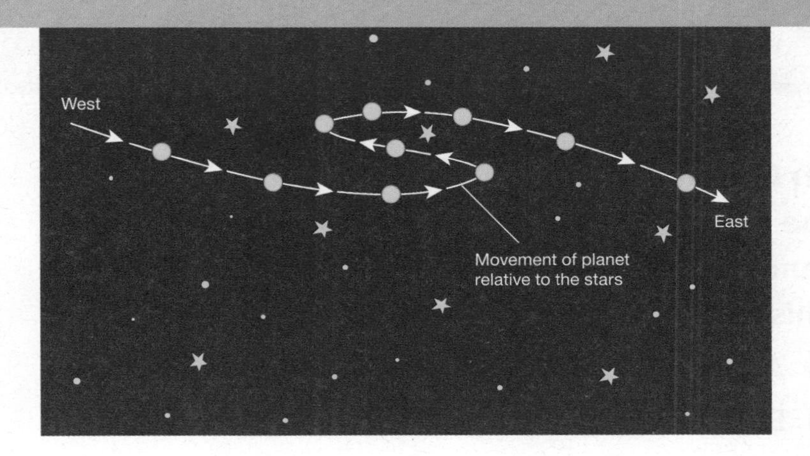

Watching the Sky

Planets and stars look like tiny dots of light. It is hard to tell them apart. If the small dot seems to twinkle, it's probably a star. Planets shine with a steady light.

Suppose you watched a planet every night for many months. You'd see that it seems to move back and forth through the stars. The stars also move. They are so far away from Earth, it is hard to detect their motion.

Lesson Review

Complete this main idea statement.

1. There are several ways to classify groups of ___________.

Complete these detail statements.

2. Imaginary pictures made of stars are ___________.
3. The ___________ is made up of billions of galaxies.
4. Fill in this graphic organizer.

Stars

The Sun and Other Stars	Seasonal Star Positions
___________	___________
___________	___________
___________	___________
___________	___________

Groups of Stars

____

____

Fill in the circle in front of the letter of the best choice.

1. The cycle of the moon's phases can be seen because

◯ A. the moon's distance from Earth changes at a predictable rate.

◯ B. the moon spins on its axis.

◯ C. the moon's axis is tilted.

◯ D. the moon revolves around Earth. S4E2b

2. Summer days in the Northern Hemisphere are hotter than winter days, because in summer

◯ A. Earth's Northern Hemisphere is tilted toward the sun.

◯ B. Earth is closer to the sun.

◯ C. the sun gets hotter.

◯ D. the sun gives off more energy. S4E2c

3. On Monday night, you see a bright object in the sky. On Thursday night, you notice the same object in a slightly different location. This tells you that the object is

◯ A. a star.

◯ B. the moon.

◯ C. an asteroid.

◯ D. a planet. S4E1b

4. Which is true of stars and planets?

◯ A. Stars move through the sky, but planets don't.

◯ B. Stars and planets both move.

◯ C. It is easier to see planets than stars because planets glow with heat.

◯ D. Stars are the brightest objects in the night sky. S4E1b

5. A star's color tells us about its

◯ A. temperature.

◯ B. size.

◯ C. rotation.

◯ D. distance from Earth. S4E1a

6. **Which statement correctly describes the outer planets?**

◯ A. The outer planets are warmer than the inner planets.

◯ B. The outer planets are smaller than the inner planets.

◯ C. The outer planets are made of gases.

◯ D. The outer planets have rocky surfaces.

S4E2d

7. **One Earth day is the time required for one**

◯ A. revolution of Earth.

◯ B. rotation of Earth.

◯ C. rotation of the sun.

◯ D. revolution of the sun.

S4E2a

8. **Which statement about the planets is true?**

◯ A. Pluto is no longer classified as a planet.

◯ B. The outer planets are solid.

◯ C. Earth is an outer planet.

◯ D. The inner planets have longer years than the outer planets.

S4E2d

9. **Suppose you see the constellation Orion. How will this constellation look if you view it from the same place 12 months from now?**

◯ A. It will look different because Earth will have orbited once around the sun.

◯ B. It will look different because stars move across the sky.

◯ C. It will look the same because stars are so far from Earth that it is hard to detect their motion.

◯ D. It will look different because Earth will have rotated on its axis.

S4E1c

10. **If scientists discovered a large, round object revolving around a star, that object would be considered a**

◯ A. moon.

◯ B. planet.

◯ C. star.

◯ D. solar system.

S4E1b

11. What are the gas giants?

◯ A. the outer planets

◯ B. comets

◯ C. stars in the Milky Way

◯ D. the sun and Venus S4E2d

12. What is the correct order of the inner planets?

◯ A. Jupiter, Saturn, Uranus, Neptune

◯ B. Mercury, Venus, Earth, Mars

◯ C. Mercury, Earth, Mars, Venus

◯ D. Jupiter, Mars, Uranus, Neptune S4E2d

13. Which separates the inner planets from the outer planets?

◯ A. Venus

◯ B. comets

◯ C. an asteroid belt

◯ D. Earth S4E2d

14. Which statement BEST explains why the sun appears to rise in the east and set in the west?

◯ A. Earth orbits the sun.

◯ B. Earth rotates on its axis.

◯ C. The moon orbits Earth.

◯ D. The sun orbits Earth. S4E2a

15. During winter in the Northern Hemisphere, which statement is true?

◯ A. The North Pole is tilted away from the sun.

◯ B. Day and night are the same length.

◯ C. The sun's rays shine more directly on the North Pole than the South Pole.

◯ D. Neither pole is tilted toward the sun. S4E2c

16. Which of the following is one way technology is used to observe distant objects in the sky?

◯ A. using spacecraft to observe the outer planets

◯ B. using a microscope to observe stars

◯ C. using binoculars to look at a bug

◯ D. using a flashlight to look at the stars S4E1d

17. Mario uses a softball to model moon phases in his darkened classroom by turning in a swivel chair placed next to a lamp. At times, Mario's body is between the lamp and the softball. At other times, the softball is between Mario and the lamp. Based on this information, which of the following statements is true?

- ◯ A. Mario's arm represents the inner planets.
- ◯ B. Mario's turning represents Earth's orbit around the sun.
- ◯ C. Mario represents the sun and the chair represents Earth.
- ◯ D. Mario represents Earth, the softball represents the moon, and the lamp represents the sun.

S4E2b

18. You are holding a model of Earth and the sun. You rotate the model of Earth once on its axis. This represents one

- ◯ A. minute.
- ◯ B. day.
- ◯ C. month.
- ◯ D. year.

S4E2a

19. Which of the following about star color is true?

- ◯ A. The oldest stars are blue.
- ◯ B. The hottest stars are blue.
- ◯ C. The newest stars are yellow.
- ◯ D. The hottest stars are red.

S4E1a

20. During January, Jupiter will suddenly become visible in the western sky. Why isn't it visible all year?

- ◯ A. because Earth rotates
- ◯ B. because Jupiter is a star, and stars change their positions in the sky
- ◯ C. because Jupiter moves across the sky from left to right
- ◯ D. because Earth and Jupiter revolve around the sun at different rates

S4E1c

The Big Idea

The sun's energy causes water and air to move in ways that result in predictable weather patterns.

On this page, record what you learn as you read the chapter.

Essential Question

What is the water cycle?

____

____

____

Essential Question

How do the oceans and the water cycle affect weather?

____

____

____

____

Essential Question

How is weather predicted?

____

____

____

____

____

____

____

____

Quick and Easy Project

Make a Rain Gauge

Procedure

1. CAUTION: **Be careful when using scissors.** Remove the cap from the bottle, and have an adult cut the top off the bottle.
2. Tape the ruler to the outside of the bottle. The zero mark should be at the bottom of the bottle.
3. Turn the cut-off bottle top over so that it will act like a funnel, and insert it in the bottle.
4. Put the rain gauge out in the open, but away from any roof edges or trees.
5. After it rains, measure the rainfall, and then empty the bottle.

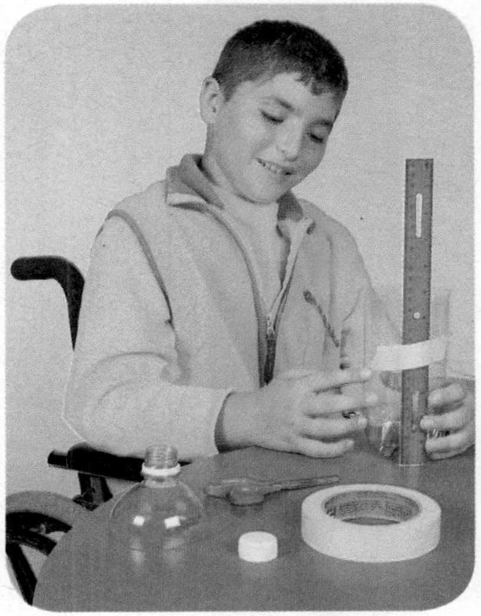

Materials:

- 1-L clear plastic bottle
- scissors
- plastic ruler
- masking tape

Draw Conclusions

1. How much rainfall did you measure?

____

2. How did measuring rainfall help you describe weather?

____

____

Independent Inquiry

Weather and the Seasons

Design an investigation in which you use various weather instruments to measure weather over the course of a year. Record data regularly. Use your data to compare measurements such as average daily temperature, wind speed and direction, and amount of precipitation among the different seasons. Then draw graphs that show how weather in your area changes over the year.

Georgia Performance Standards

S4E3a Demonstrate how water changes states from solid (ice) to liquid (water) to gas (water vapor/steam) and changes from gas to liquid to solid.

S4E3b Identify the temperatures at which water becomes a solid and at which water becomes a gas.

S4E3d Explain the water cycle (evaporation, condensation, and precipitation).

Vocabulary Activity

1. On the picture of the water cycle, circle the arrow that represents evaporation.
2. When water becomes a gas, it is called ________ ____________.

Student eBook
www.hspscience.com

Lesson 1

VOCABULARY
water cycle
water vapor
evaporation
condensation

What Is the Water Cycle?

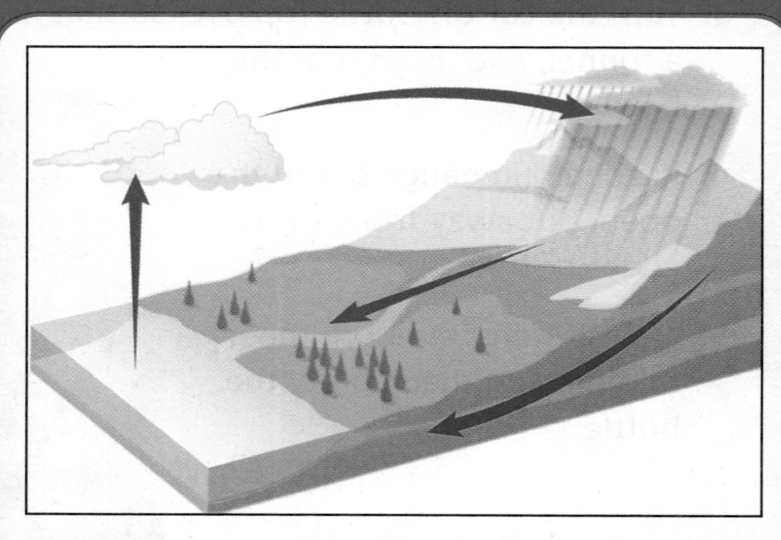

As the sun heats Earth's liquid water, it becomes a gas. Later, this gas turns back to a liquid. It falls from the sky to the ground. This happens over and over. This is the **water cycle**.

The water in the pot boiled. The liquid water became a gas called **water vapor**.

Have you ever seen a puddle dry up? **Evaporation** causes the liquid water to become water vapor. You can't see it, but it's there in the air.

Some mornings you see dew on plants. Water vapor in the air turned to liquid water. That is **condensation** at work.

Insta-Lab

How Much Water?

1. Fill a 1-L container with water.
2. Add 4 drops of food coloring.
3. Put 28 mL of the water into a small, clear container.
4. From the small container, put 7 mL of the water into another small, clear container.
5. Observe how much water is in each container. The remaining water in the 1-L container represents all the salt water on Earth. The water in the first small container represents all the fresh water on Earth. The water in the second small container represents all the liquid fresh water on Earth.
6. How important do you think it is to protect our freshwater resources?

____

____

✓ Quick Check

The main idea of these two pages is Water covers most of Earth's surface. Details tell more about the main idea. Underline two details about water.

1. Why can't you drink water from the ocean?

____

2. Why do people need fresh water?

____

The Water Planet

Water covers most of Earth's surface. This is why many people call Earth the water planet. Most of Earth's water is in oceans. That water is salty. You can't drink it. You can't use it to grow plants. People need fresh water to drink, wash, and grow plants.

Most of Earth's surface is covered with water.

Most people can't use the fresh water in glaciers. The ice sheets are frozen. They are also too far from where most people live.

Most of Earth's freshwater is frozen. It is in big sheets of ice. These are called *glaciers*. Earth does have some fresh water in rivers and lakes. There is even fresh water underground. People pump it up to Earth's surface.

✓ **Quick Check**

1. Where is most of Earth's fresh water?

____

2. Why can't most people use the fresh water in glaciers?

____

____

3. Where can people find fresh water that they can use?

____

4. Where is most of Earth's water found?

____

Quick Check

The main idea of these two pages is Water is always moving. Details tell more about the main idea. Circle three details about water movement.

1. Look at the picture on this page. Put an **X** on an arrow that shows water changing from a liquid to a gas and moving up into the air.

2. What is water called when it changes from a liquid into a gas?

____

The Water Cycle

Water is always moving. It goes from Earth's surface into the air. Then it comes back down again. This is called the **water cycle**. Here's how it works.

First, the sun warms Earth's water. Some water changes from a liquid to a gas. This gas, called **water vapor**, moves up into the air.

In the air, the water vapor cools. It changes back to liquid water. In clouds, water droplets join to make larger drops. When the drops become heavy enough, they fall back to Earth.

The drops may fall into an ocean, a lake, or a river. They may soak into the ground. Or they may be warmed by the sun and become water vapor again.

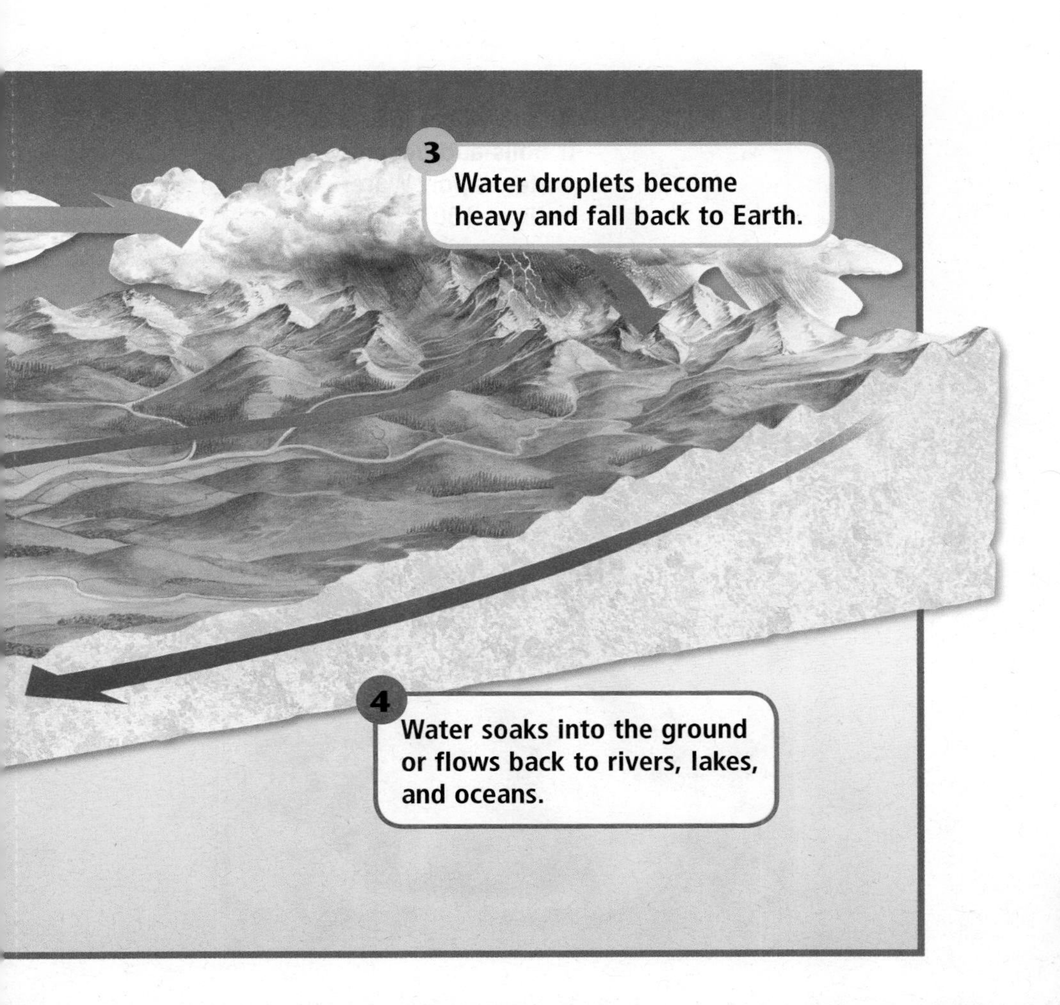

✓ Quick Check

1. When water vapor in the air changes back to a liquid, what does it form?

____

2. When do the drops of water fall back to Earth?

____

3. What could happen to a drop that falls back to Earth?

____

____

4. Why is this process called a cycle?

____

5. Use the space below to draw a picture of the cycle, using the labels *evaporation*, *condensation*, and *water vapor*.

Quick Check

The main idea of these two pages is Water changes from a liquid to a gas and a gas to a liquid. Details tell more about the main idea. Circle two details about water changing from one state to another.

1. Water freezes at __________ °C.

2. Water boils at __________ °C.

3. Tell where water goes when it evaporates.

____

____

Evaporation

Evaporation is what happens when a liquid changes into a gas. Liquid water evaporates. It changes to water vapor. Water vapor is invisible. It mixes with other gases in the air.

As the water is heated, it boils and becomes water vapor. Water boils at 100°C.

Condensation

Condensation is what happens when a gas changes into a liquid. Water vapor cools. It changes to a liquid. Water droplets mix with bits of dust in the air. This makes clouds. Very high clouds may have tiny bits of ice. Why? The water drops froze.

Liquid water freezes at 0°C.

Lesson Review

Complete this main idea statement.

1. The water cycle explains how

____

____

Complete these detail statements.

2. ____________________ is the changing of a liquid to a gas.

3. ____________________ is the changing of a gas to a liquid.

4. Fill in this graphic organizer.

Main Idea: Water on Earth is found in many different places.

salt water	fresh water
Found in: ______________ ______________	Found as ice in: ______________ Found as liquid water in: ______________ ______________

Georgia Performance Standards

S4E3c Investigate how clouds are formed.

S4E3e Investigate different forms of precipitation and sky conditions (rain, snow, sleet, hail, clouds, and fog).

Vocabulary Activity

Write the meanings of the vocabulary words on the lines below.

1. Current

____

2. Humidity

____

3. Weather

____

4. Precipitation

____

Lesson 2

How Do the Oceans and the Water Cycle Affect Weather?

VOCABULARY
current
weather
humidity
precipitation

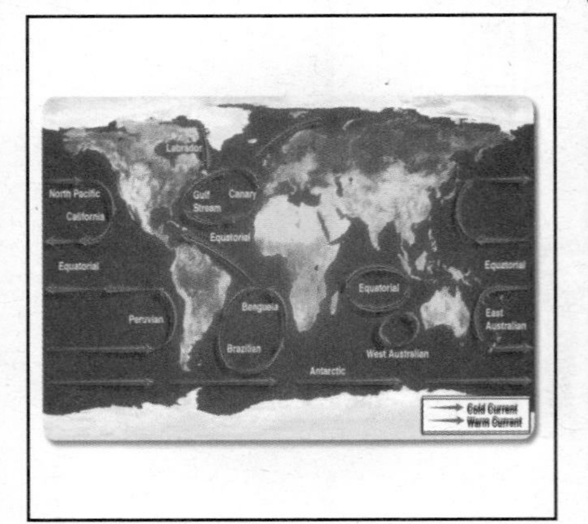

A **current** is a stream of water. Most currents flow through an ocean. The Gulf Stream is a current in the Atlantic Ocean. It flows off the coast of Georgia.

Weather is the condition of the air at a certain place and time. Much of what we call weather is part of the water cycle.

The amount of water vapor in the air is called **humidity**. If the humidity is high, the air feels sticky and damp. If the humidity is low, the air feels very dry.

Rain and snow fall from the sky. These are two different kinds of **precipitation**. That's any form of water that falls back to Earth.

Insta-Lab

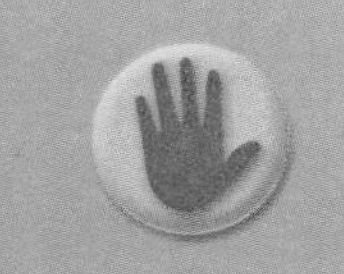

Kitchen El Niño

1. Model what happens in the oceans when warm and cold water mix.
2. Fill a large container with warm water (approximately 80°F).
3. Fill a small cup with very cold water (approximately 40°F), and add a few drops of food coloring.
4. Use tongs to lower the cup straight down below the surface of the warm water. Observe what happens.
5. How might this explain why clouds form over ocean water?

____

____

____

✓ Quick Check

Focus Skill A cause makes something happen. An effect is what happens. What is the cause for more clouds forming over warm water than over cold water?

____

____

What is the effect of wind pushing on the surface water of the oceans?

____

1. On the map at the bottom of the page, circle one current that always has warm water, and make an **X** on two currents that always have cold water.
2. Tell what causes a current.

____

____

The Oceans Affect Weather

The sun heats ocean water. But the ocean does not heat up evenly. Some parts are warmer than other parts. At the same time, wind pushes the water on the surface. This creates a **current**, or stream of water. Some currents carry warmer water toward colder parts of the ocean.

Warm water evaporates more quickly than cool water. More clouds form over warm water. Those clouds bring rain. The ocean's temperature will affect how much it rains in a certain area.

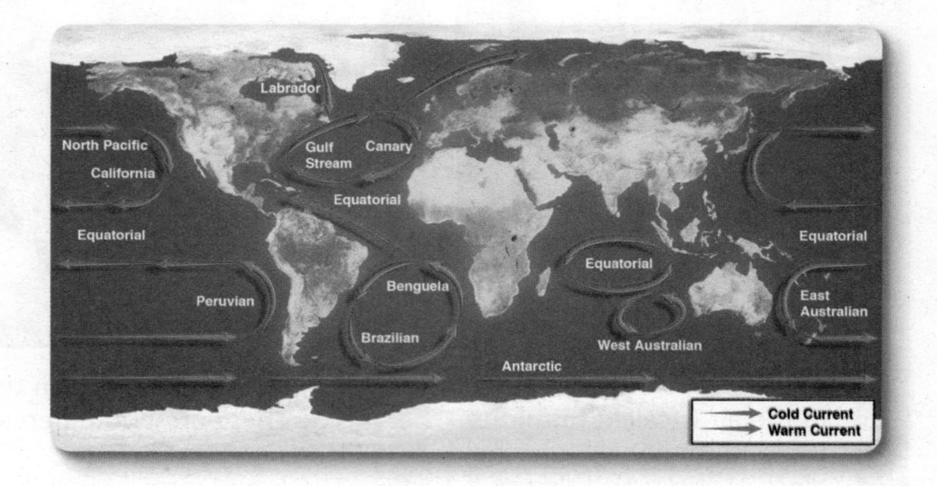

▲ **The Gulf Stream is a current. It moves warm water from the Gulf of Mexico to the Georgia coast.**

Weather Patterns and the Water Cycle

Thanks to winds, water vapor moves. It goes from warmer to cooler places. If it didn't, all the water would just fall right back down where it started.

Water vapor that forms over a warm ocean can move a long way. It carries heat as it moves. When it condenses, it releases heat energy. This is how the water cycle brings changes in the weather.

▲ Winds move warm water vapor to cooler areas.

1. What causes water vapor to move?

____

2. In the picture at the bottom of the page, circle the arrow that shows how wind moves water vapor. Write to explain where winds move warm water vapor.

____

3. Tell how the water cycle affects weather.

____

____

____

✓ Quick Check

A cause makes something happen. An effect is what happens. Circle the cause of fog forming. Underline an effect of warm air moving upward.

1. Why do clouds form in air that has a lot of water vapor?

____

____

____

2. Tell what causes clouds to form.

____

____

____

Clouds

Weather is the condition of the air at a certain place and time. For example, *humid* weather means that the air contains a lot of water vapor. **Humidity** is the amount of water vapor in the air. Clouds form in air with a lot of water vapor.

As warm air moves upward, it cools. Some water vapor condenses on bits of dust. Soon, a cloud forms. Sometimes water condenses near Earth's surface and forms fog.

Tell what causes clouds to form.

Cumulus Clouds

Cumulus (KYOO•myuh•luhs) clouds are puffy. They usually mean good weather. But if they are gray at the bottom, rain may come.

Stratus Clouds

Stratus (STRAT•uhs) clouds look long and flat. They are low in the sky. They usually mean some rain or snow is coming.

Cirrus Clouds

Cirrus (SIR•uhs) clouds form high in the sky, where the air is very cold. They are made mostly of ice crystals.

Kinds of Clouds

Precipitation

Water from clouds returns to Earth as **precipitation**. The water may fall as rain, snow, sleet, or hail. If water vapor turns directly into ice, it snows. If water passes through freezing-cold air, it falls as sleet or hail.

▲ **In warm weather, water falls as rain.**

Lesson Review

Complete the following cause-and-effect statements.

1. An ocean ____________ is caused by wind pushing water on the surface.
2. Warm water ______________ faster. This causes more clouds to form over warm water.
3. Water vapor carries ______________ from warmer to cooler places.
4. Fill in this graphic organizer.

Cause →	Effect
____________________ ____________________	It rains.
____________________ ____________________ ____________________	It snows.
Water vapor near Earth's surface condenses into small water droplets.	______________ .

Georgia Performance Standards

S4E4a Identify weather instruments and explain how each is used in gathering weather data and making forecasts (thermometer, rain gauge, barometer, wind vane, anemometer).

S4E4b Using a weather map, identify the fronts, temperature, and precipitation and use the information to interpret the weather conditions.

S4E4c Use observations and records of weather conditions to predict weather patterns throughout the year.

S4E4d Differentiate between weather and climate.

Vocabulary Activity

Use the vocabulary words to fill in the table.

Vocabulary Term	Definition
	a weather tool that tells how fast the wind is blowing
	where two air masses meet
	a weather tool that measures air pressure
	a section of air
	the study of weather
	what the weather is like over many years

GO online Student eBook www.hspscience.com

Lesson 3

How Is Weather Predicted?

VOCABULARY
meteorology
barometer
anemometer
hygrometer
air mass
front
climate

Meteorology is the study of weather. Lots of kids study weather. They use weather tools. The kids want to know what the weather will be like tomorrow!

A **barometer** is a weather tool. It measures air pressure. The amount of pressure may rise or fall.

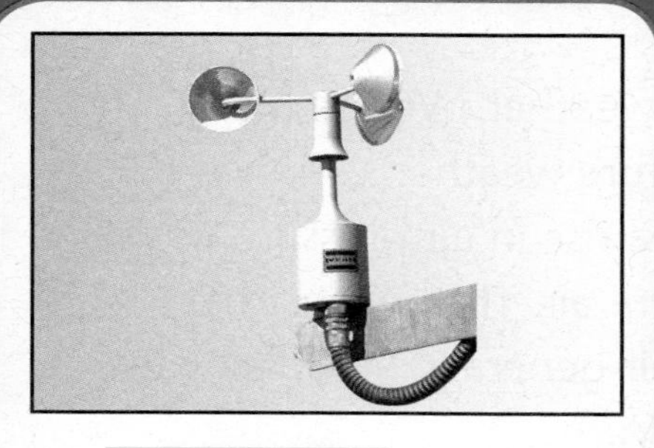

An **anemometer** is another weather tool. It tells how fast the wind is blowing.

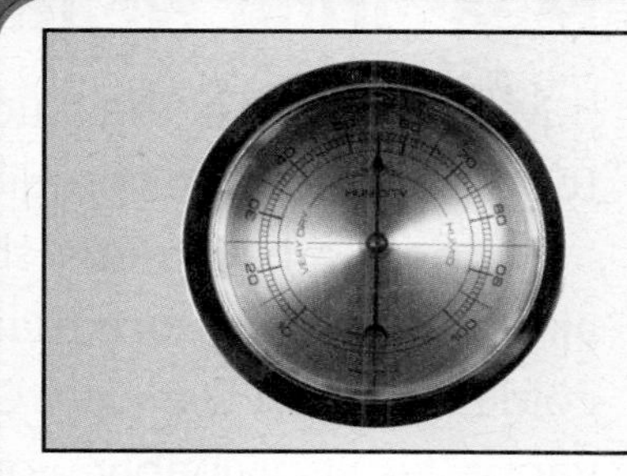

A **hygrometer** tells how much water vapor is in the air.

An **air mass** is a section of air. All the air in it has the same temperature and humidity.

A **front** is where two air masses meet.

Climate is what the weather is like over many years. Different parts of Georgia have different climates.

Insta-Lab

Shining a Light on Climate

1. Model how sunlight strikes Earth between the equator and the poles.
2. Tape a sheet of paper to a book.
3. Shine a lamp or a flashlight straight at the paper.

4. Hold the light still, and tilt up one end of the book. The part of the book closest to the light models the equator. The other end of the book models the pole.
5. Write what you observe.

____

____

6. Use what you observed to explain why the equator is warmer than other areas of Earth.

____

Quick Check

The main idea on these two pages is There are different ways to measure weather. Details tell more about the main idea. Underline two details about how to measure weather.

1. Why doesn't it usually rain when a barometer shows a rise in air pressure?

2. How can rising air pressure help people predict the temperature of the air?

Measuring Weather

Meteorology is the study of weather. We use many tools to forecast, or predict, future weather.

Suppose a barometer shows a rise in air pressure. Cold air has more pressure than warm air. This means that colder air is on the way. Cold air generally has less water vapor. So, it probably won't rain.

	Weather Instrument	Measures
	Thermometer	This tool measures air temperature.
	Barometer	This tool measures air pressure.
	Anemometer	This tool measures wind speed.
	Hygrometer	This tool measures the amount of water vapor in the air.

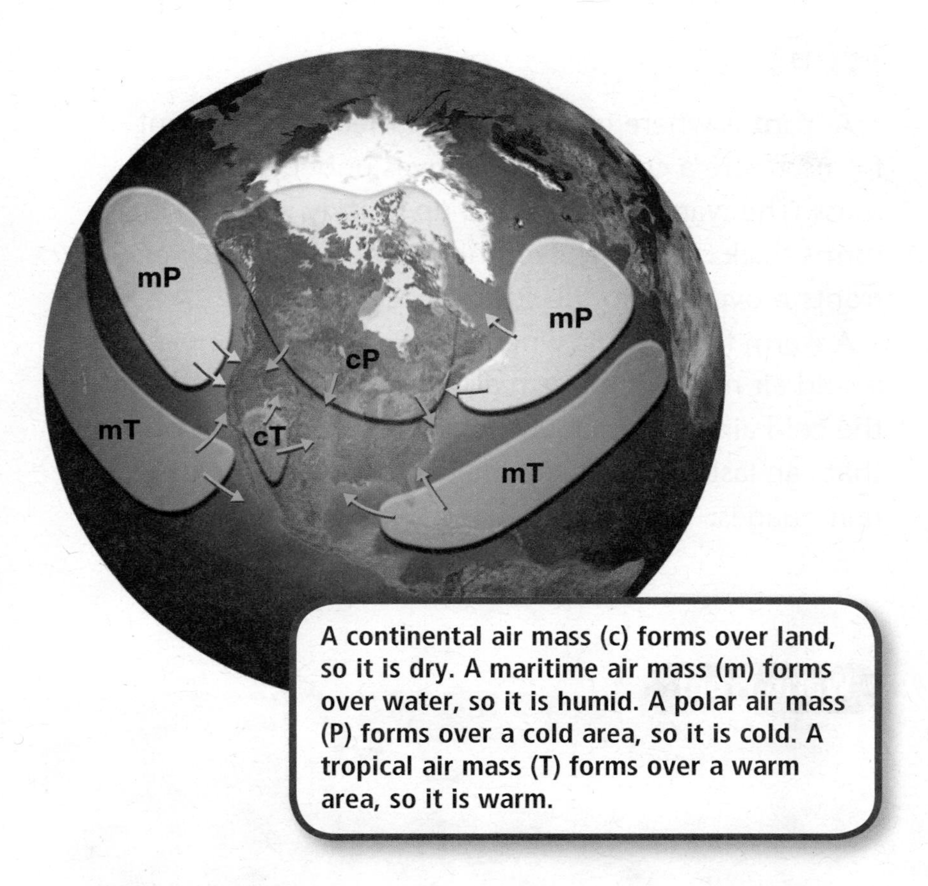

A continental air mass (c) forms over land, so it is dry. A maritime air mass (m) forms over water, so it is humid. A polar air mass (P) forms over a cold area, so it is cold. A tropical air mass (T) forms over a warm area, so it is warm.

Air Masses

The sun does not heat the air evenly. Instead, air moves in bunches, or masses. All of the air in an **air mass** has the same temperature and humidity. An air mass that forms near the equator is humid and warm. Air masses near the poles are dry and cold.

As air masses move, they create winds. Air masses generally move from west to east across the United States. Wind vanes can measure wind direction.

Quick Check

1. Why are continental air masses dry?

____

____

2. What causes wind?

____

____

3. Which weather instrument measures wind direction?

____

4. Where do hot, humid air masses form?

____

✓ Quick Check

Focus Skill The main idea on these two pages is There are two kinds of fronts and each forms where two air masses meet. Details tell more about the main idea. Underline two details about fronts.

1. Which type of front brings storms that pass quickly?

____

2. What kinds of clouds form as a warm front passes through?

____

3. What kind of weather does a warm front bring?

____

____

Fronts

A **front** is where two air masses meet. A cold front forms where a cold air mass moves under a warm air mass. The warm air is pushed up quickly. It cools and forms thick clouds. They bring heavy rain or snow. Cold fronts move fast, so the storms pass quickly.

A warm front forms where a warm air mass moves over a cold air mass. The warm air slowly slides up and over the cold air. Stratus clouds form. They bring rain or snow that can last for hours. Rainfall amounts are measured by rain guages.

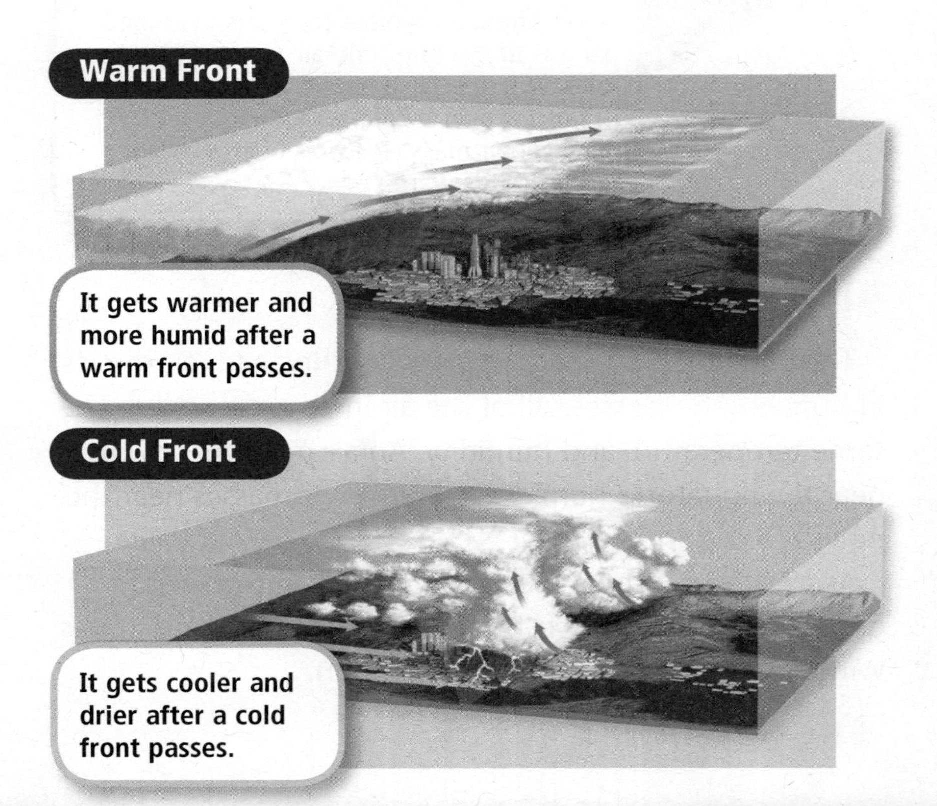

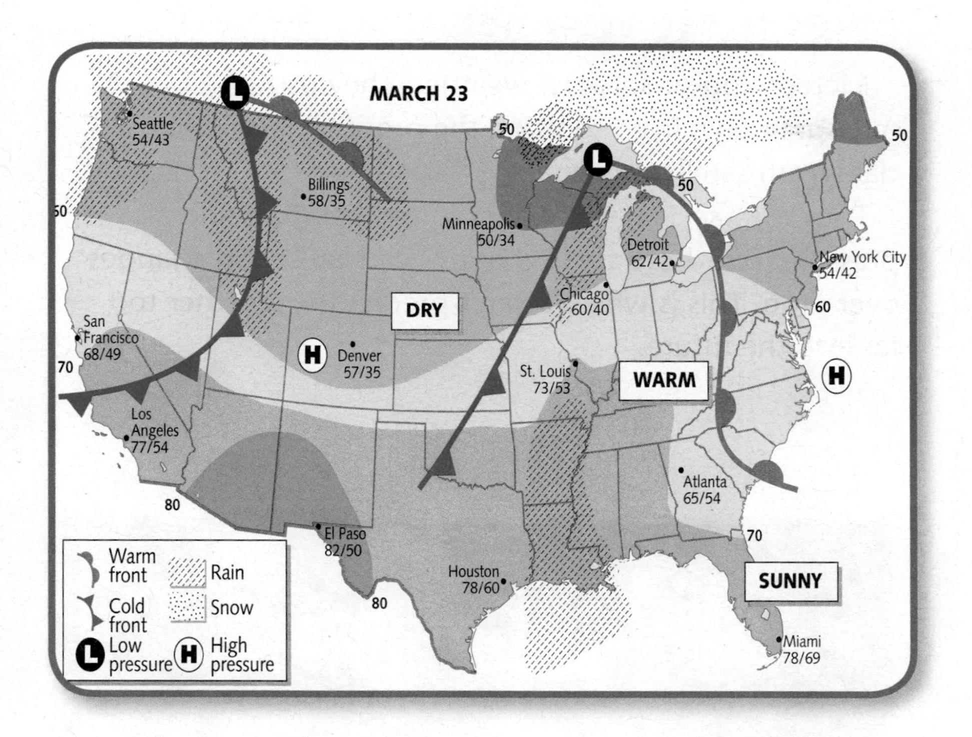

Weather Maps

Weather maps use symbols. The map key shows what each symbol means.

Look at the map. Find a red line with half circles. It shows a warm front. A blue line with triangles shows a cold front. A weather map may also list high and low temperatures. It may show wind speed and direction. And it may show air pressure.

Quick Check

1. Look at the map on this page. Circle the city that had rain that day.
2. Draw a box around the city where the temperature was 65 degrees.
3. Which city is close to an area of high pressure?

   ____

4. Which city on the map had the highest temperature that day?

   ____

5. How are fronts shown on a weather map?

   ____

   ____

   ____

Quick Check

The main idea on these two pages is Meteorologists forecast weather. Details tell more about the main idea. Underline two details about how meteorologist forecast weather.

1. Look at the picture on this page. How can satellite pictures like this help meteorologists forecast weather?

____

____

____

2. Why is it not possible to predict weather far into the future?

____

____

Forecasting Weather

Meteorologists forecast weather. They use measurements from tools on the ground. They also use data from satellites in space.

They may see small changes in temperature and air pressure. But these small changes can cause big changes over time. This is why we can't predict the weather too far into the future.

This picture from space shows a hurricane heading for Georgia.

Weather Patterns and Climates

Weather in most places follows a pattern. It may generally be cool in the morning and warm in the afternoon. It all depends on how much sunlight you get. Many places have different weather in winter, spring, summer, and fall. **Climate** is a place's average weather over many years.

▲ **Earth is tilted on its axis. Different amounts of sunlight reach different places. This causes the seasons.**

Lesson Review

Complete this main idea statement.

1. Scientists use information from weather tools to ___________ future weather.

Complete this detail statement.

2. As an ___________ moves, it produces wind.

3. Explain the difference between weather and climate.

____

____

____

4. Fill in this graphic organizer.

Main Idea: Measurements taken with weather instruments can be used to forecast weather.	
___________	___________
___________	

Fill in the circle in front of the letter of the best choice.

1. In the water cycle, water in the ocean evaporates. Water vapor moves up into the atmosphere, where it cools. What happens to the water vapor when it cools?

- ◯ A. The water vapor melts.
- ◯ B. The water vapor changes to hail.
- ◯ C. The water vapor evaporates.
- ◯ D. The water vapor condenses.

S4E3c

Use the weather map below to answer question 2.

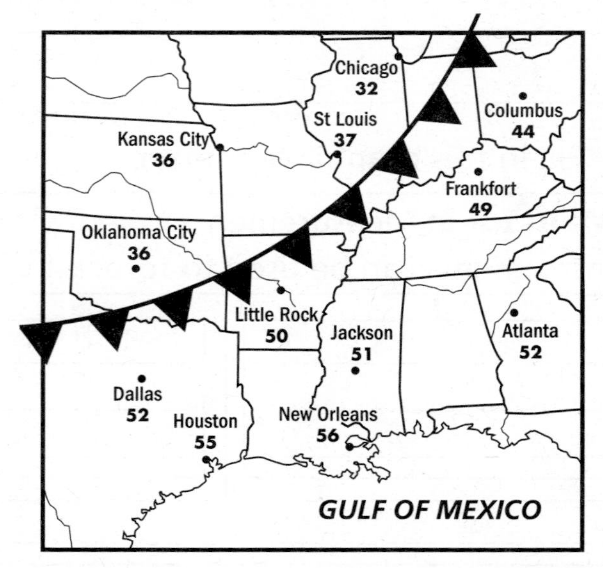

2. What is the weather in Georgia likely to be like in the next few days?

- ◯ A. A warm front will soon pass over, so weather will be cooler and drier.
- ◯ B. A warm front will soon pass over, so weather will be warmer and more humid.
- ◯ C. A cold front will pass over, so weather will be cooler and drier.
- ◯ D. A cold front will pass over, so weather will be warmer and more humid.

S4E4b

3. The weather is very cold, and the air is filled with water vapor. These are good conditions for

- ◯ A. sleet.
- ◯ B. snow.
- ◯ C. hail.
- ◯ D. rain

S4E3e

Use the diagram below to answer questions 4 and 5.

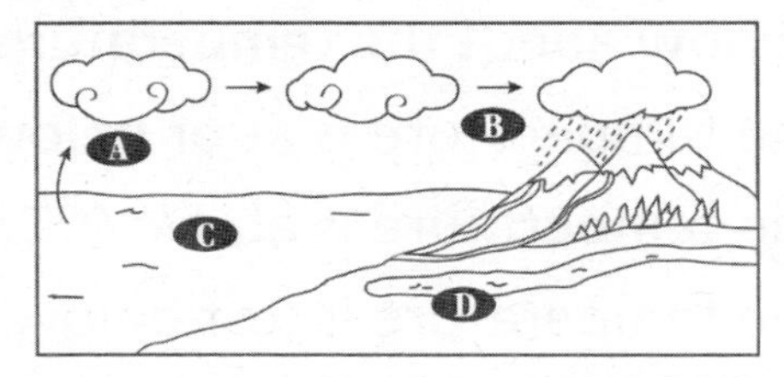

4. What is happening at the arrow labeled B?

- ◯ A. Water is evaporating.
- ◯ B. Water is freezing.
- ◯ C. Water is condensing.
- ◯ D. Water is falling as precipitation.

S4E3d

5. How does water at arrow A form a cloud?

- ◯ A. The water freezes.
- ◯ B. The water evaporates.
- ◯ C. The water condenses.
- ◯ D. The water precipitates.

S4E3d

6. If water vapor turns directly to ice, it

- ◯ A. rains.
- ◯ B. sleets.
- ◯ C. snows.
- ◯ D. hails.

S4E3e

7. Before a thunderstorm, the air pressure drops. Which instrument could help you decide if you need to carry an umbrella?

- ◯ A. balance
- ◯ B. scale
- ◯ C. barometer
- ◯ D. windsock

S4E4a

8. You look out your window and see hail. What caused this type of precipitation?

- ◯ A. Water vapor condensed on the ground.
- ◯ B. Water vapor condensed and froze at ground level.
- ◯ C. Water vapor turned directly into ice crystals.
- ◯ D. Water droplets passed through very cold air.

S4E3e

9. On a weather map, cold fronts are indicated by

- ◯ A. the letter H.
- ◯ B. the letter L.
- ◯ C. a blue line with triangles.
- ◯ D. a red line with half circles.

S4E4b

10. **If a winter wind is blowing from the north, the weather is likely to get colder. Which instrument could help you decide which coat to wear to school?**

- ◯ A. anemometer
- ◯ B. barometer
- ◯ C. hygrometer
- ◯ D. weather vane

S4E4a

11. **Fog can form when**

- ◯ A. clouds descend to Earth's surface.
- ◯ B. air masses collide along a front.
- ◯ C. the ground loses heat more slowly than the air does.
- ◯ D. water vapor in humid air directly above the ground condenses.

S4E3e

12. **A cup of ice sits in the sun. The ice melts. What change of state will occur NEXT?**

- ◯ A. from solid to solid
- ◯ B. from liquid to solid
- ◯ C. from liquid to gas
- ◯ D. from gas to liquid

S4E3a

13. **Suppose the surface of a lake is frozen. What do you know about the temperature of the ice?**

- ◯ A. The temperature is at or below 0°C.
- ◯ B. The temperature is above 0°C.
- ◯ C. The temperature is just below 100°C.
- ◯ D. The temperature is above 100°C.

S4E3b

14. **On a warm morning, the meteorologist states that water vapor is condensing near the ground. This tells you that you may see**

- ◯ A. snow.
- ◯ B. sleet.
- ◯ C. fog.
- ◯ D. rain.

S4E3e

15. **Even when it is warm outside, a strong wind can make it feel much colder. Which instrument could help you decide if you need to wear a coat to school?**

- ◯ A. thermometer
- ◯ B. anemometer
- ◯ C. weather vane
- ◯ D. barometer

S4E4a

Use the drawing below to answer question 16.

16. The sun's heat will cause water in the puddle to change from

- ◯ A. a gas to a liquid.
- ◯ B. a liquid to a gas.
- ◯ C. A solid to a liquid.
- ◯ D. a liquid to a solid.

S4E3a

17. In the water cycle, what happens before water condenses to form clouds?

- ◯ A. Water falls as rain.
- ◯ B. Water evaporates.
- ◯ C. Water vapor changes to a gas.
- ◯ D. Water dissolves salt in the ocean.

S4E3d

18. At which temperature does water boil?

- ◯ A. 0°F
- ◯ B. 32°F
- ◯ C. 100°C
- ◯ D. 212°C

S4E3b

19. Which statement is TRUE?

- ◯ A. Climate is the average of the conditions of the atmosphere over many years.
- ◯ B. Weather is the average of the conditions of the atmosphere over many years.
- ◯ C. Climate can change quickly.
- ◯ D. When the Northern Hemisphere receives direct sunlight, the climate gets colder.

S4E4d

20. Where is most of Earth's water located?

- ◯ A. in lakes
- ◯ B. in rivers
- ◯ C. in the oceans
- ◯ D. in the ground

S4E3d

Sound and Light

Sound and light are forms of energy that interact with matter in ways that cause us to see and hear the things around us.

On this page, record what you learn as you read the chapter.

Essential Question

What is sound?

____

____

____

____

Essential Question

What is light?

____

____

____

____

Essential Question

How do objects bend light?

____

____

____

____

____

____

Quick and Easy Project

Kaleidoscope

Materials:

- colored markers
- sheet of white paper
- 2 rectangular mirrors

Procedure

1. Use colored markers to make dots, stars, and other small shapes on a sheet of white paper. Your marks should be in an area about 3 cm square.
2. Stand one mirror on its short edge near your marks. Then look down into the mirror so that you can see the reflection of your marks.
3. Stand a second mirror at an angle to the first one, and look at the reflections in the two mirrors. Change the angle between the mirrors, and observe how the reflections change.

Draw Conclusions

1. Describe the reflections you saw first with the two mirrors.

____

____

2. How did changing the angle between the mirrors affect the reflections?

____

____

Independent Inquiry

Stringed instruments make sounds when the strings are plucked, bowed, or struck, causing them to vibrate. Use an empty tissue box and rubber bands of different sizes to make a stringed instrument. You will need two pencils to slip under the rubber bands, one on each side of the opening in the box. Use your stringed instrument to test variables that affect the kind of sound produced.

Georgia Performance Standards

S4P2a Investigate how sound is produced.

S4P2b Recognize the conditions that cause pitch to vary.

Vocabulary Activity

Sound is a form of energy produced by vibrations. See how differences in vibrations change the sound.

Fill in the chart. Look at the pictures for help.

	the number of vibrations per second of an object
	how loud a sound is
	how high or low a sound is
	a back-and-forth motion

Student eBook
www.hspscience.com

Lesson 1

What Is Sound?

VOCABULARY
vibration
volume
pitch
frequency

Vibration is a back-and-forth motion. If you pluck a guitar string, the string moves back and forth very quickly. It *vibrates*.

The **volume** of a sound is how loud it is. Some sounds are very loud. They have a greater volume than softer sounds do.

A sound can be high or low. A sound's **pitch** is how high or low it is.

Objects vibrate at different speeds. When one object vibrates faster than another, there are more vibrations per second. The number of vibrations per second is the **frequency**.

Insta-Lab

Playing the Glasses

1. Fill several glasses with different levels of water.
2. Tap gently on each glass with a pencil.
3. Add water to the glasses.
4. Tap on each glass again.

How does adding water to the glass change the pitch of the sounds you made?

___

___

___

✓ Quick Check

The main idea of these two pages is Sound is a form of energy that happens when something vibrates. Details tell more about the main idea. Underline two details about how sound energy can be produced.

1. How do vibrations cause sound?

____

____

____

2. In the space below, draw three other items that make sound.

Sound Energy

Sound is a form of energy that happens when something vibrates. **Vibration** is a quick back-and-forth movement of matter.

A sound starts when an object vibrates. The vibrations cause the air around the object to vibrate. The vibration of the air makes many of the sounds you hear.

Musical instruments make sounds by making different things vibrate. A drum vibrates when it is hit, and that causes the sound.

A jackhammer is so loud that it can damage your hearing.

Violins have strings. The player plucks the strings or draws a bow across them. That causes the strings to vibrate, which makes a sound. Flutes make a sound when the air inside them vibrates.

Sounds can be loud or soft. The loudness of a sound is its **volume**. Some sounds have more energy than others. Those with more energy have greater volume.

A drum vibrates when it is hit.

✓ Quick Check

1. Fill in the table below to tell how instruments make sounds.

drum	
	Air inside it vibrates, causing sound.
violin	

2. What causes some sounds to be louder than others?

____

____

3. What is vibration?

____

____

Quick Check

The main idea of these two pages is Sound travels through the air as waves. Details tell more about the main idea. Underline two details about the way sound travels.

1. Describe how sound moves through the air.

____

____

____

2. What is pitch?

____

3. What is frequency?

____

4. Circle the wave in both images on this page.

Sound Waves

Sound travels through the air as waves. When an object vibrates, it pushes the air around it. The air is compressed, or squeezed together. The compressed air then pushes the air next to it. That passes the compression along. This makes a sound wave.

A sound's **pitch** is how high or low it is. Pitch comes from the speed of the vibrations. The number of vibrations per second is the **frequency**. Small objects usually vibrate faster than large ones do.

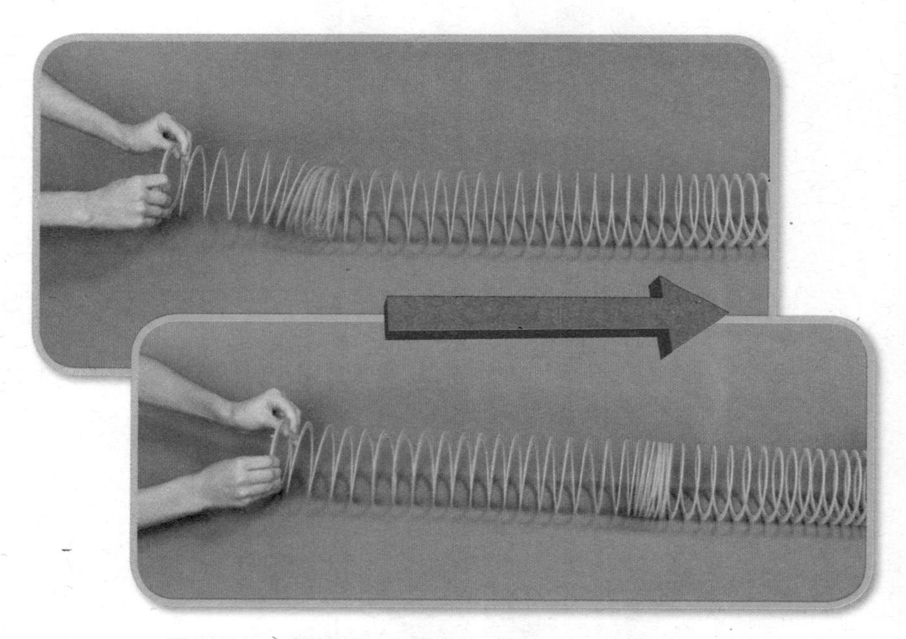

▲ **These springs show how a sound wave moves. The compressed part moves along the spring, but the spring stays in place.**

An object that vibrates quickly has a high frequency. A sound with a high frequency has a high pitch. An object that vibrates slowly has a low frequency. A sound with a low frequency has a low pitch.

Sound waves move away from a vibrating object in all directions. When a sound wave hits something, energy is absorbed. Soft surfaces absorb more sound than hard ones do. Sound waves that hit a large, hard surface bounce back and may create an echo.

Sound waves bouncing off a large, hard surface can cause an echo.

✓ Quick Check

1. Does an object that vibrates quickly have a high or a low pitch?

____

2. In which direction does sound move away from an object?

____

3. What causes echoes?

____

____

____

4. Circle the echo in the image.

The main idea of these two pages is Sound waves of many different frequencies can carry energy a long distance. Details tell more about the main idea. Underline two details about how sound energy travels.

1. What happens to air as sound travels through it?

____

____

2. What must matter be able to do to carry sound?

____

Sound Transmission

Sound waves can carry energy a long distance. The energy travels to a new place, but the matter that carries it does not. Sound waves move through the air because air particles vibrate in place.

Air is not the only matter that carries sound. Any kind of matter can vibrate and carry sound. Sound travels at different speeds depending on the matter that carries it.

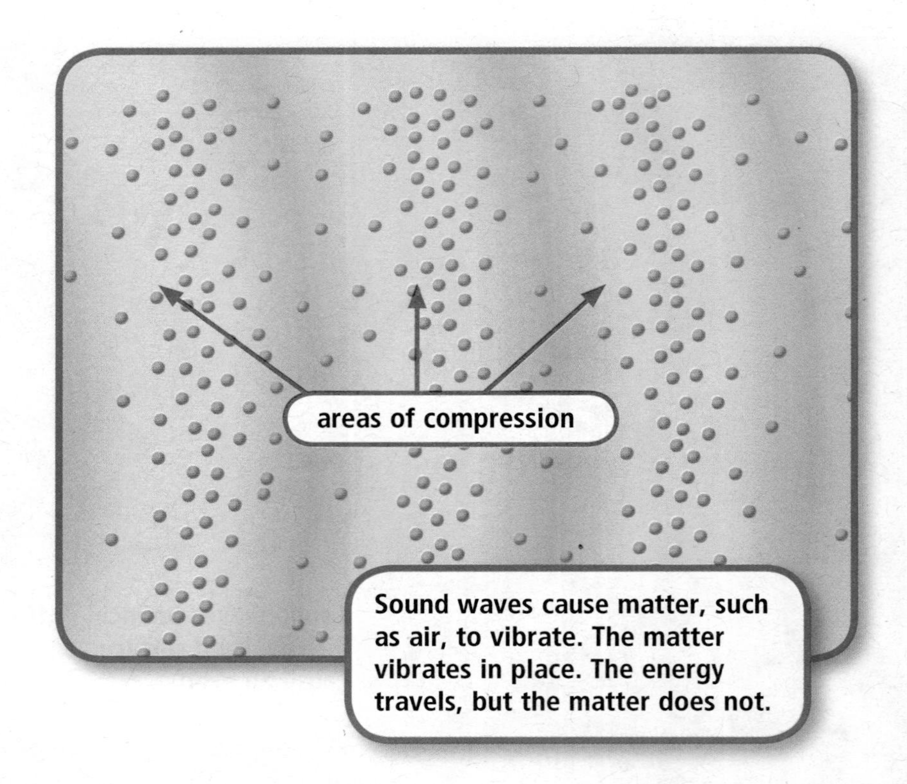

Sound waves cause matter, such as air, to vibrate. The matter vibrates in place. The energy travels, but the matter does not.

Animals and Sound

People can hear sounds over a wide range of frequencies. But some animals can hear sounds of much higher or lower frequencies. Bats make high-frequency sounds that help them fly at night. Echoes of the sounds tell bats where things are.

Other animals can make and hear sounds of lower frequencies than people can. Elephants make low-frequency sounds that travel long distances. They use the sounds to communicate with each other.

Elephants communicate using low-pitched sounds.

Lesson Review

Complete this main idea statement.

1. Sound travels through air or other matter as ______________.

Complete these detail statements.

2. The loudness of a sound is called its __________.
3. An object that vibrates quickly has a high frequency and a ______________ pitch.
4. Fill in this graphic organizer.

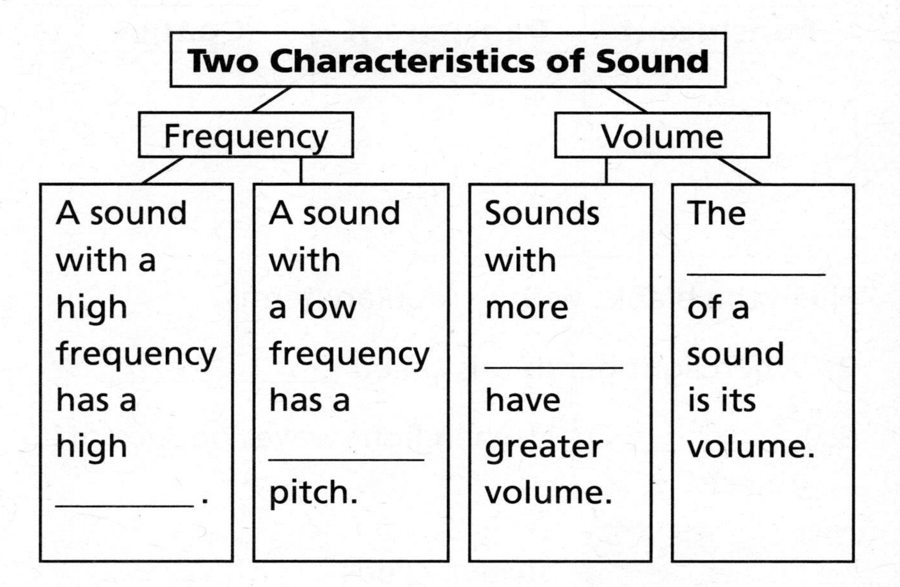

Georgia Performance Standards

S4P1a Identify materials that are transparent, opaque, and translucent.

S4P1b Investigate the reflection of light using a mirror and a light source.

Vocabulary Activity

In this lesson, you will learn about light and how it travels.

1. In the table below, give two examples that fit the vocabulary terms.

Translucent	Transparent	Opaque

Fill in the blanks with vocabulary terms.

2. When light bends it is called ______________.

3. ______________ is when light waves bounce off objects.

Student eBook
www.hspscience.com

Lesson 2

What Is Light?

VOCABULARY
reflection
refraction
translucent
transparent
opaque

Light waves bounce off objects. That is **reflection**. Reflected light allows you to see an object.

Light waves can move from one kind of matter to another. When that happens, the speed of the light changes, which causes the light to bend. That bending is called **refraction**.

An object that is **translucent** lets only some light pass through it. You cannot see very clearly through a translucent object.

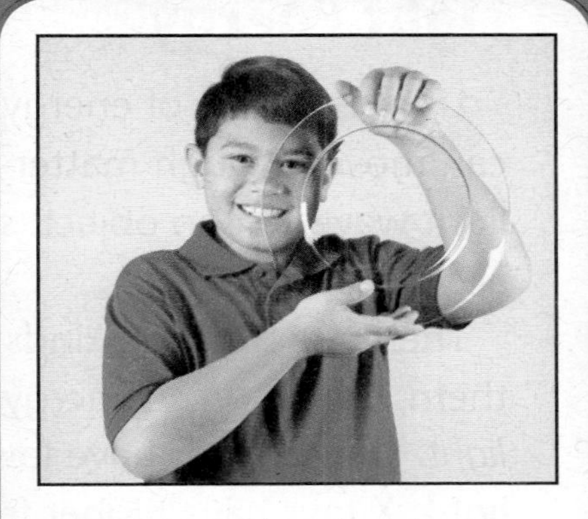

An object that is **transparent** lets light pass through it. It is easy to see clearly through a transparent object.

An object that is **opaque** does not let light pass through it. You cannot see through an opaque object.

Insta-Lab

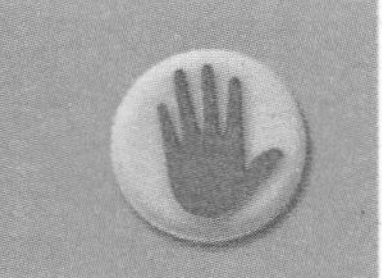

Look At and Through

1. Hold sheets of plastic wrap, aluminum foil, and wax paper up to a light.
2. Which sheet is opaque?

   ____

   How do you know?

   ____

3. Which sheet is translucent? ________________

   How do you know?

   ____

   ____

4. Which sheet is transparent? ________________

   How do you know?

   ____

   ____

✓ Quick Check

The main idea of these two pages is Light is a form of energy that travels in waves. Details tell more about the main idea. Underline two details about light traveling in waves.

1. Sound must move through matter. How is light different?

____

____

2. How are radio waves different from visible light?

____

____

3. Circle the higher frequency waves in the image on this page.

Light Energy

Light is a form of energy that travels in waves. Light can move through matter or through empty space. When light waves hit an object, some of the light energy may be absorbed.

There are different kinds of light waves. We cannot see them, though. The energy we can see is called *visible light*. Radio waves have lower frequencies than visible light. X rays have higher frequencies than visible light. The higher the frequency, the more energy a wave has.

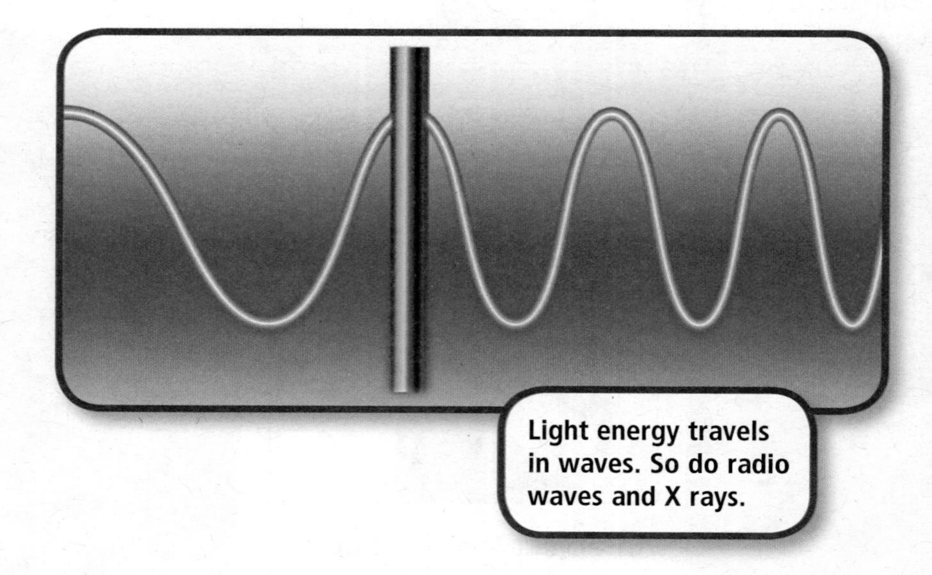

Light energy travels in waves. So do radio waves and X rays.

Light Waves

Light travels in waves. They are not like sound waves, though, because they are not compression waves. Instead, light waves move up and down, forming an S shape. This kind of wave is called a *transverse* wave.

Transverse waves carry only energy from one place to another. If the waves travel through matter, the matter moves up and down. But the matter itself does not move forward or backward.

Since light waves do not need matter to move, they can travel through empty space. Light moves very fast. As far as scientists know, there is nothing faster than light.

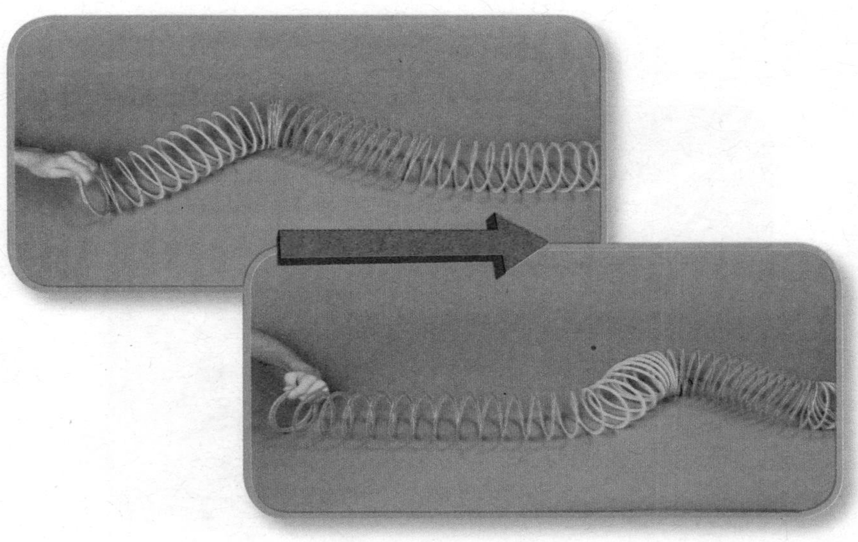

▲ **Light waves move up and down through matter. The matter does not move forward or backward.**

✓ Quick Check

1. What kind of shape does a transverse wave make?

____

2. When transverse waves travel through matter, does the matter move?____________

   If so, describe the way the matter moves.

____

____

3. Circle the sentence that tells why light can move through empty space.

4. How do light waves move?

____

Quick Check

The main idea of these two pages is When light hits an object, the path of the light can change. Details tell more about the main idea. Underline two details about how the path of light can change.

1. Why do you think dark-colored objects appear dark when light hits them?

____

____

____

2. What can happen when light hits an object?

____

____

Absorption and Reflection

When light hits an object, the path of the light can change. Some objects absorb light, while others make light bounce off. Still others let light pass through.

Dark objects absorb more light than light-colored ones do. Objects also reflect light. **Reflection** is the bouncing of light from a surface. Reflected light is what lets you see an object.

Reflected light usually scatters in all directions. A smooth surface reflects light in a pattern. You see the pattern as as an image on the smooth surface. That is what happens when you see yourself in a mirror.

A mirror reflects light in a pattern.

Refraction

Sometimes light moves from one kind of matter to another. That changes the speed of the light, which makes the light bend. **Refraction** is the bending of light as it moves between different kinds of matter.

An object that is partly in water can appear to be broken in two pieces. Light waves that travel through the air follow one path to your eyes. Light waves that travel through the water follow another path. The result is that you see the object in two parts.

▲ When a beam of light enters water straight on, the light does not bend, but it does slow down.

▲ When a beam of light enters water at an angle, the light slows down and bends.

✓ Quick Check

1. What happens when light moves from one kind of matter to another?

____

2. What causes an object that is partly in water to appear broken?

____

____

____

3. Look at the pictures on these two pages. Circle the picture that shows light being reflected.

4. List details about refraction.

____

____

____

____

____

✓ Quick Check

The main idea of these two pages is Some materials allow different amounts of the light that hits them to pass through. Details tell more about the main idea. Underline two details about how light passes through objects.

1. Insert the vocabulary terms in the table below to show how much light each object allows to pass through it.

Objects that allow the most light to pass through them	Objects that allow some light to pass through them	Objects that don't allow any light to pass through them

2. What makes an object transparent?

____

____

Translucent, Transparent, and Opaque

Some materials allow only part of the light that hits them to pass through. They are **translucent**. These materials absorb or scatter some light. For that reason, you can't see through them clearly. A frosted light bulb is translucent. It lets light through, but you cannot see what is inside the bulb.

Other materials allow most of the light that hits them to pass through. They are **transparent**. You can see clearly through transparent objects, such as most glass and water.

Transparent marbles

Translucent marbles

Some materials do not let any light pass through them. They are **opaque**. Opaque objects, such as wood and metals, absorb more light than translucent objects do.

Opaque stones

Lesson Review

Complete this main idea statement.

1. Light is a form of ______________ that travels through matter or through empty space in waves.

Complete the detail statement.

2. A smooth surface ______________ light in a pattern, which you see as an image.

3. Fill in this graphic organizer.

Light Waves Change When They Hit Objects

Opaque objects ______________ some light and reflect the rest.	Smooth surfaces ______________ a lot of light.

Transparent objects let light pass through, although they make the light __________ , or refract.

Georgia Performance Standard

S4P1c Identify the physical attributes of a convex lens, a concave lens, and a prism and where each is used.

Vocabulary Activity

In the boxes below, draw a picture of each vocabulary term.

Concave Lens	Convex Lens

Student eBook
www.hspscience.com

Lesson 3

How Do Objects Bend Light?

VOCABULARY
concave lens
convex lens

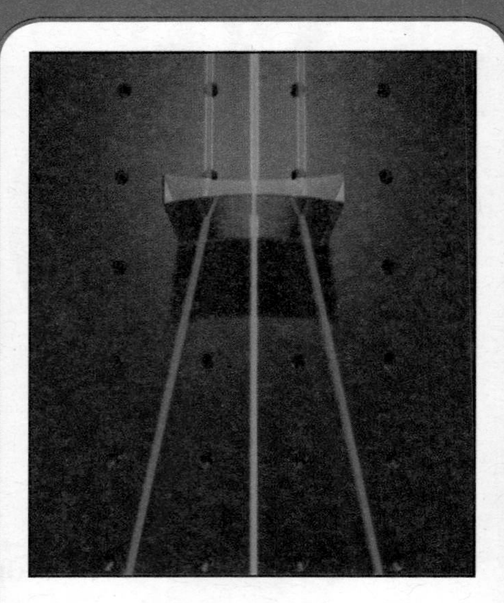

A *lens* is a curved, transparent object that can bend light. A **concave lens** is thin in the middle and thick at the edges.

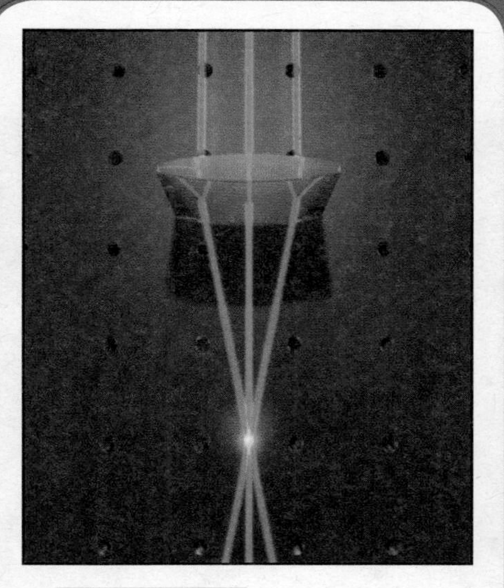

A **convex lens** is different from a concave lens. A convex lens is thick in the middle and thin at the edges.

Insta-Lab

Water Lens

1. Use water and a test tube to model one type of lens.
2. Fill a test tube completely with water.
3. Close the filled test tube with a stopper.
4. Hold the test tube over some writing.
5. How does your "water lens" change what you see?

____

6. What kind of lens is it?

____

7. How does the shape of the test tube compare to the shape of that type of lens?

____

____

✓ Quick Check

Focus Skill A cause makes something happen. An effect is what happens. Circle the cause of a white light being separated. Underline an effect that is caused by white light entering a prism.

1. How does a prism separate white light into the separate colors?

____

____

____

2. Fill in the chart. Tell the effect of each cause.

Cause	Effect
Light enters the prism.	
Each color refracts a little differently	
Light leaves a prism.	

3. Where might a prism be used? Why?

____

Light and Color

A beam of white light contains all the colors in a rainbow. A prism separates the colors of light because it changes the direction of light waves.

When light enters a prism, the waves change direction. Each color refracts, or bends, a little differently than the others, so the colors take different paths. The light also changes direction when it leaves the prism, bending the waves again. This allows you to see each color. Prisms are also used in cameras and binoculars to direct light.

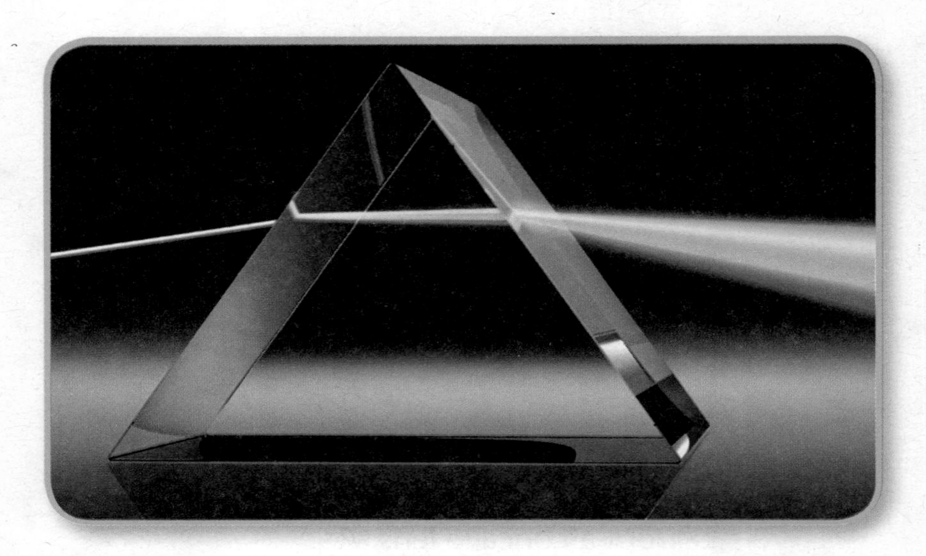

▲ A prism bends light. That separates the colors.

◀ **Other objects refract light just as prisms do.**

Opaque objects appear to be colored because they reflect certain waves of light. A red object reflects red light waves and absorbs other colors. An object that reflects all the colors together appears to be white. An object that absorbs all the colors appears to be black.

Small drops of water in the air can create a rainbow. The drops have the same effect on light that a prism does. They bend light, too. As light waves enter and leave each drop, the light changes direction. The result is that the colors separate.

A drop of water can act like a prism.

✓ Quick Check

1. Other than a prism, what are two other things that can separate white light into colors?

___

2. What makes a blue shirt appear blue?

___

___

___

3. Why would an object that reflects all the colors together appear white?

___

___

___

4. In the image of a water drop, circle the places where light is bent.

✓ Quick Check

Focus Skill A cause makes something happen. An effect is what happens. Circle the cause of a convex lens making things look bigger than they are. Underline an effect that is caused by light waves bending toward the end of a lens.

1. Which type of lens makes things look bigger than they are?

____

2. Can you think of some items that use convex lenses?

____

____

3. What effect does a convex lens have on light?

____

4. What kind of lens do you have in your eyes?

____

Lenses

A lens is transparent and curved, and it refracts, or bends, light. A **convex lens** is thick in the middle and thinner at the edges. Light waves that go through the lens bend toward the thickest part, which is the middle. A convex lens makes things look bigger than they are.

A **concave lens** is thick at the edges and thinner in the middle. Light waves that enter it bend toward the ends, where the lens is thickest. A concave lens makes things look smaller than they are.

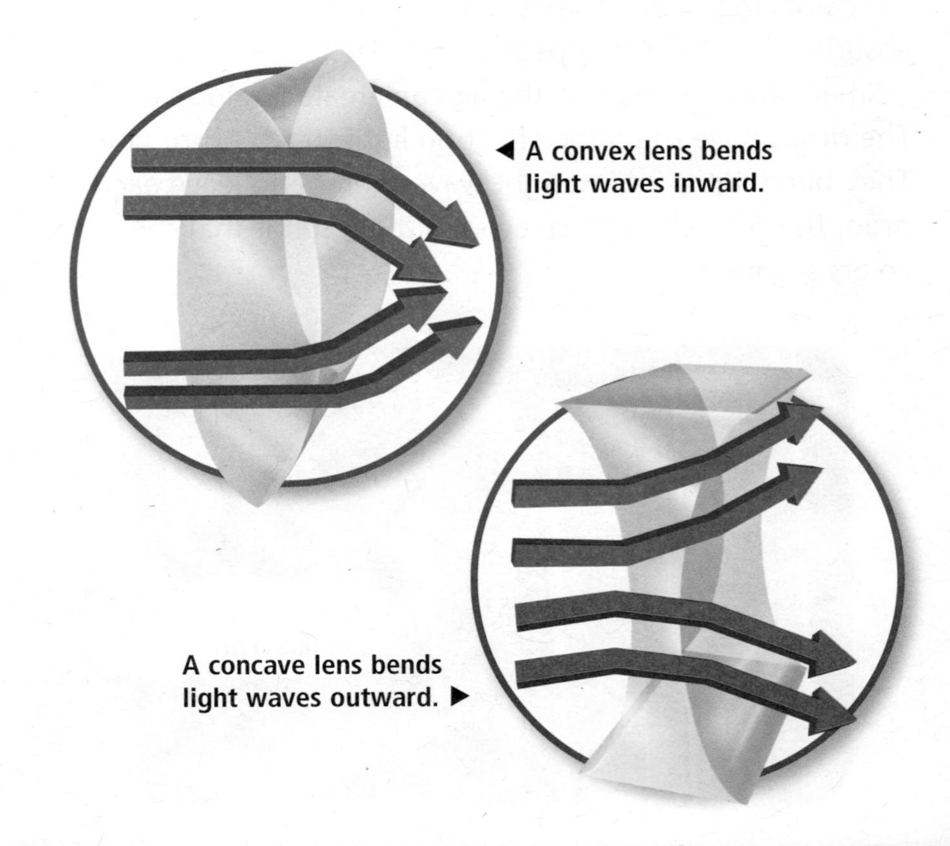

◄ A convex lens bends light waves inward.

A concave lens bends light waves outward. ►

Vision

Each of your eyes contains a lens. The lens is convex, and it focuses light on the retina. The retina is at the back of your eye.

Some people have trouble seeing things that are far away. Concave lenses, either in glasses or contact lenses, can solve that problem. The lenses focus the light at the correct point in the eye. Other people are not able to see things that are up close. Convex lenses also change the point where light is focused, making close objects clearer.

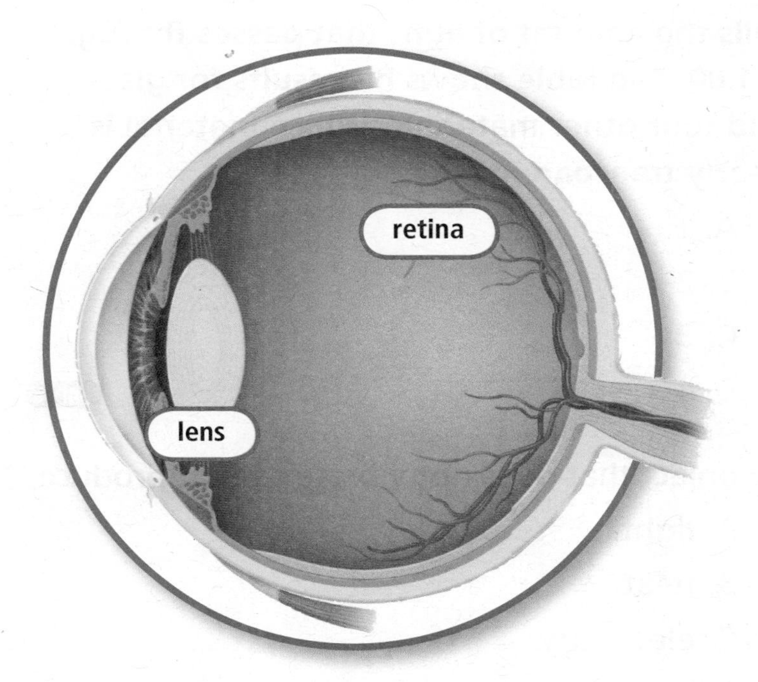

▲ **The convex lens in each of your eyes focuses light.**

Lesson Review

Complete the following cause-and-effect statements.

1. Prisms, drops of water, and lenses all cause light waves to ______________.

2. Opaque objects appear to be colored because they ______________ a certain color of light.

3. Fill in this graphic organizer.

Cause	Effect
Light passes through a prism.	______________
______________	Light rays bend inward. Objects look bigger or an image is formed.
Light passes through a concave lens.	______________

CRCT Practice

Fill in the circle in front of the letter of the best choice.

1. A homeowner wants to install a skylight to get as much light as possible into a dark upstairs hallway. What kind of glass should she use?

- ◯ A. mirrored
- ◯ B. opaque
- ◯ C. translucent
- ◯ D. transparent

S4P1a

Use the graph below to answer question 2.

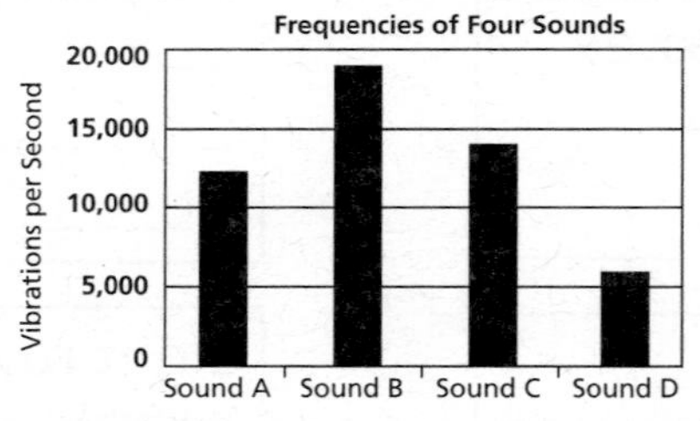

2. The graph shows the number of vibrations per second for four different sounds. Order the sounds from highest pitch to lowest pitch.

- ◯ A. B, C, A, D
- ◯ B. A, B, D, C
- ◯ C. D, A, C, B
- ◯ D. A, C, D, B

S4P2b

Use the table below to answer question 3.

Material	Amount of Light that Passes Through
Glass	1.00
1	0.10
2	0.25
3	0.95
4	0.03

3. Sam uses a light meter to measure the amount of light that passes through different materials. He first measures clear glass. He calls the amount of light that passes through it 1.00. The table shows his results for glass and four other materials. Which material is nearly transparent?

- ◯ A. 1
- ◯ B. 2
- ◯ C. 3
- ◯ D. 4

S4P1a

4. An object that does not vibrate cannot produce

- ◯ A. light.
- ◯ B. heat.
- ◯ C. electricity.
- ◯ D. sound.

S4P2a

5. All of these windows face the same pine tree. Which window is opaque?

- A.
- B.
- C.
- D.

S4P1a

6. You stand in front of a window. A friend waves to you from outside. Why can you see your friend clearly?

- A. The glass is a prism.
- B. The glass is translucent.
- C. The glass is opaque.
- D. The glass is transparent.

S4P1a

7. What happens if you shine a flashlight on a mirror in a dark room?

- A. The light causes the mirror to get warm.
- B. The mirror reflects the light back to you.
- C. The light passes through the mirror.
- D. The mirror absorbs all the light.

S4P1b

8. Which of these is a concave lens?

- A.
- B.
- C.
- D.

S4P1c

9. A musician plucks two guitar strings. One makes a lower-pitched sound. What is true of that string?

- A. It is shorter than the other string.
- B. It vibrates more slowly than the other string.
- C. It produces more energy than the other string.
- D. It moves faster than the other string.

S4P2b

10. A lens that is thicker in the middle than at the edges is

- A. convex.
- B. concave.
- C. light-reflecting.
- D. prismatic.

S4P1c

11. Marcia wraps a book in wax paper. She can see the book's title, but the letters look fuzzy. Why?

- ◯ A. The wax paper is transparent.
- ◯ B. The wax paper is translucent.
- ◯ C. The wax paper is concave.
- ◯ D. The wax paper is convex.

Use the diagram below to answer question12.

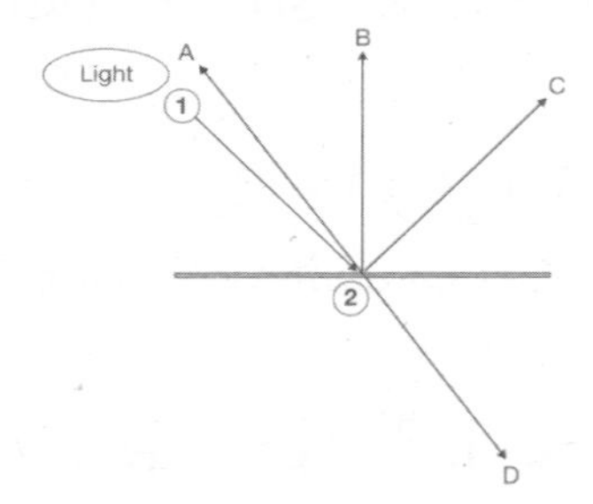

12. Light travels from point 1. It strikes a surface at point 2. If it is reflected in the direction of C, then which of the following is true?

- ◯ A. The surface is transparent.
- ◯ B. The surface is a mirror.
- ◯ C. The surface is convex.
- ◯ D. The surface is translucent.

S4P1b

13. Why can you see an arch of colors in the sky on a rainy day?

- ◯ A. Raindrops act like convex lenses.
- ◯ B. Raindrops act like concave lenses.
- ◯ C. Raindrops act like prisms.
- ◯ D. Raindrops act like rainbows.

14. What produces sound?

- ◯ A. applying a force to an object
- ◯ B. causing an object to vibrate
- ◯ C. shining light on an object
- ◯ D. heating an object

15. What happens to the sound an object makes when the speed of vibrations decreases?

- ◯ A. Its volume increases.
- ◯ B. Its pitch rises.
- ◯ C. Its pitch becomes lower.
- ◯ D. Its volume decreases.

S4P2b

16. A microscope makes objects look larger than they actually are. Which kind of lens does it contain?

- ◯ A. opaque
- ◯ B. translucent
- ◯ C. concave
- ◯ D. convex

S4P1c

17. You stand in front of your friend. Your friend hides behind you. Why can't people see your friend?

- ◯ A. You are transparent.
- ◯ B. You are translucent.
- ◯ C. You are opaque.
- ◯ D. You are concave.

S4P1a

18. Which of the following should you use to demonstrate how light reflects off a smooth surface?

- ◯ A. a convex lens
- ◯ B. a mirror
- ◯ C. wax paper
- ◯ D. a concave lens

S4P1b

19. What happens to light when it strikes a prism?

- ◯ A. It is heated.
- ◯ B. It is absorbed.
- ◯ C. It is refracted.
- ◯ D. It is straightened.

S4P1a

20. Which object bends light rays, bringing them closer together?

◯ A. ◯ B. ◯ C. ◯ D.

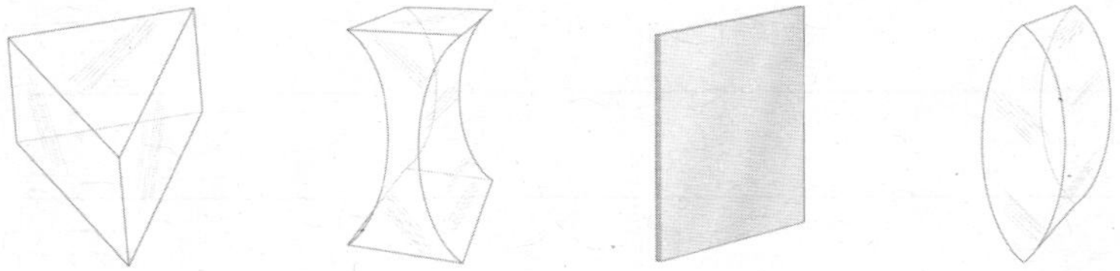

S4P1c

Forces and Motion

Forces act on and change the motion of objects in measurable ways.

On this page, record what you learn as you read the chapter.

Essential Question

How is motion described and measured?

____

____

____

____

Essential Question

How does gravity affect motion?

____

____

____

____

Essential Question

How do forces affect motion?

____

____

____

____

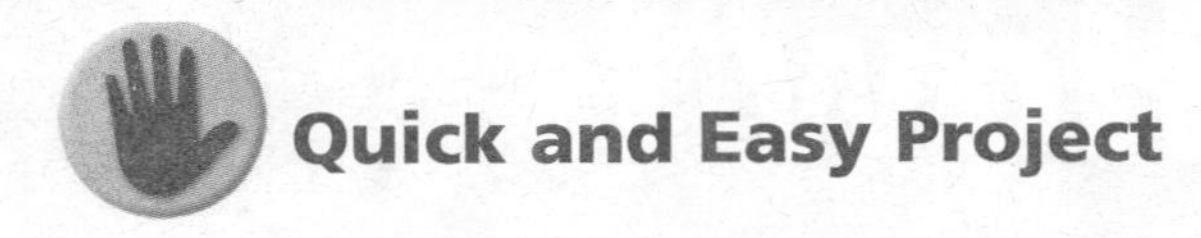

Motion and Frames of Reference

Materials:
- wagon
- round ball, such as a soccer ball

Procedure

1. Work in a group of three students. Have Student A sit in the wagon and hold the ball. Have Student B pull the wagon at a constant velocity. Student C should stand about 6 m away.
2. While the wagon is moving, have Student A toss the ball upward about 2 m and catch it. All students should observe the motion of the ball.
3. Have Students A and B repeat the motions so that all of you can verify your observations. Ask each student if the ball appears to be going straight up and down.

Draw Conclusions

Discuss each person's observations. How are the observations affected by each person's frame of reference?

____

____

____

Independent Inquiry

Change of Reference

Switch roles with your group members. Be sure each of you has a chance to observe from both places when the ball is tossed. Now change one variable. See if you get different results. You might try changing the way the ball is tossed or the location of the observer. Write what you observe.

____

____

Georgia Performance Standard

S4P3 Students will demonstrate the relationship between the application of a force and the resulting change in position and motion on an object.

Vocabulary Activity

Motion can be measured and described. In this lesson, you will learn how to describe position and changes in position. Fill in the blanks below to show how the vocabulary terms are related.

______________ is the location of an object. When an object's ______________ changes, that is called ______________. ______________ is how far something moves in a certain amount of time.

Go online Student eBook www.hspscience.com

Lesson 1

VOCABULARY
position
motion
speed

How Is Motion Described and Measured?

Position is the location of an object. It is where an object is. Some things change position. Some do not. This girl has changed her position many times.

Motion is the change of position of an object. Running, walking, going up and down, and swinging from side to side are kinds of motion.

Speed is how far something moves over a certain amount of time.

Insta-Lab

Fast Walk, Slow Walk

1. Mark off a distance of at least 10 m.
2. Walk that distance in 10 seconds.
3. Walk the same distance again. This time walk it in 15 seconds.
4. How do you determine how quickly you need to walk?

____

____

____

____

✓ Quick Check

Focus Skill You compare how things are alike. You contrast how things are different. Compare an object that changes position with one that doesn't.

____

____

Contrast the same two objects.

____

____

____

1. Circle the objects that are in motion in the picture.
2. Describe your position in the classroom. Tell where you sit in the classroom and what else is positioned near you.

____

____

Changing Position

Position is the location of an object. It is where something is. Everything has a position. Your nose is in the middle of your face. You may be at a desk in the front of the room.

Some positions do not change. Your nose stays where it is. But, if you go to the door and open it, your position changes. The position of the door changes. You and the door are in motion. **Motion** is a change of position.

◀ **Children in motion**

Look at the pictures. What can you tell about the shore and the boats? The shore is not moving. So, if you compare the positions of the boats to the shore in each picture, you can tell they have moved. You are using the shore as a *frame of reference.*

Compare the positions of the boats in the pictures. How can you tell the boats are moving? ▼

✓ Quick Check

1. Look at the pictures on this page. How can you tell the boats are moving?

____

____

____

2. When you look at the boats and are able to tell they have moved, what is your frame of reference?

____

3. How does having a frame of reference help you compare how much the boats have moved?

____

____

____

Quick Check

You compare how things are alike. You contrast how things are different. Compare distance and time.

____

____

Contrast distance and time.

____

____

1. If two students race, which student has a greater speed?

____

2. What two things do you need to measure in order to find the speed of an object?

____

Measuring Motion

How fast can you run? If you run faster than a friend, you have a greater speed. **Speed** is how far something moves over a certain amount of time.

You can find the speed of an object. You need to measure two things: distance and time. Distance is the change in an object's position. Time is how long it takes the object to move that distance.

If you know the distance, you can compare times to find out speed. ▼

Here is a way to find speed. Suppose you are on a car trip. You traveled 150 kilometers in 2 hours. Here's how to find the speed:

distance ÷ time = speed

150 km ÷ 2 hr = 75 km per hour

This means that in one hour you traveled 75 kilometers.

▼ A radar gun measures the speed of moving cars.

On the road, speed is measured in miles per hour. ▼

✓ Quick Check

1. Circle the words on this page that show how to use time and distance to find speed.
2. If you traveled 200 kilometers in 2 hours, what is your speed?

 ____

3. Which travels faster—a car going 60 miles in one hour or a car going 120 miles in two hours?

 ____

4. In the space below, write down a distance and a time. Trade books with another student and use the space to find the speed for his or her distance and time.

✓ Quick Check

Focus Skill You compare how things are alike. You contrast how things are different. Compare average speed and speed from moment to moment.

___

___

Contrast average speed and speed from moment to moment.

___

___

___

1. How do you find the average speed of an object?

___

2. How is a bar graph different from a data table?

___

___

___

Comparing Speeds

When you divide distance by time, you are really finding an *average* speed. But speed may change from moment to moment. Suppose you are watching a horse race between Tom and Jerry. So far, Tom's average speed is 55 kilometers per hour. Jerry's average speed is 60 kilometers per hour. You might guess that Jerry is moving faster than Tom. But, you must watch and compare their positions as they change during the race. If Tom is catching up to Jerry, then Tom is running faster than Jerry at that moment.

The horse in front may not be the fastest. ▶

Displaying Data About Speed

It is often helpful to display data about speed in tables. A table organizes data in columns and rows. Labels describe the data.

You can also use graphs to display speeds. A bar graph is probably best, because bar graphs are used to display data in categories. The bar graph below compares the average speeds of humans and other animals.

The longest bar represents the fastest speed. ▼

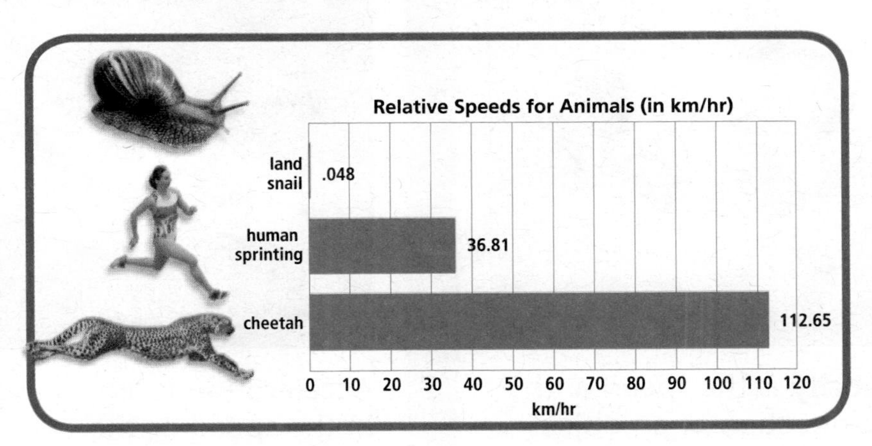

Lesson Review

Complete these compare and contrast statements.

1. Distance and ______________ are both measurements that you need to find speed.
2. Running and jumping are both kinds of ______________.
3. Fill in this graphic organizer.

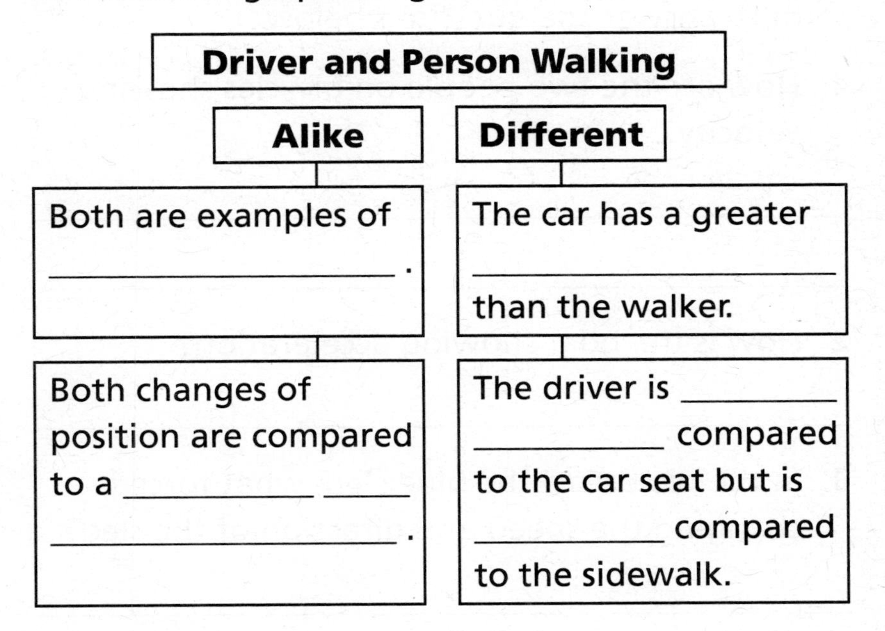

Driver and Person Walking

Alike	**Different**
Both are examples of ______________ .	The car has a greater ______________ than the walker.
Both changes of position are compared to a ______________ ______ ______________ .	The driver is ______________ ______________ compared to the car seat but is ______________ compared to the sidewalk.

Georgia Performance Standards

S4P3b Using different size objects, observe how force affects speed and motion.

S4P3c Explain what happens to the speed or direction of an object when a greater force than the initial one is applied.

Vocabulary Activity

In this lesson, you will learn about how forces affect motion. Use the picture for each vocabulary term to answer the questions below.

1. How are the two people on bicycles showing velocity?

____

____

2. How is the boat showing acceleration?

____

3. In the picture of the blue sled, what force is changing the speed and direction of the sled?

____

Student eBook
www.hspscience.com

Lesson 2

How Do Forces Affect Motion?

VOCABULARY
velocity
acceleration
force
inertia

Velocity is the speed and the direction of an object.

Acceleration is any change in the speed or the direction of an object in motion.

A **force** is a push or a pull. Forces change motion. The dog's force, or pull, moves the sled.

Inertia is a property of matter that keeps an object from changing speed or direction unless a force acts on it.

Insta-Lab

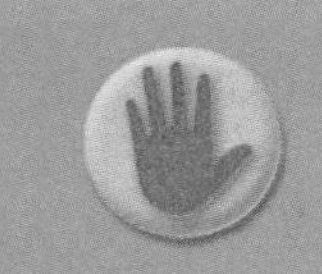

Spring-Scale Follow the Leader

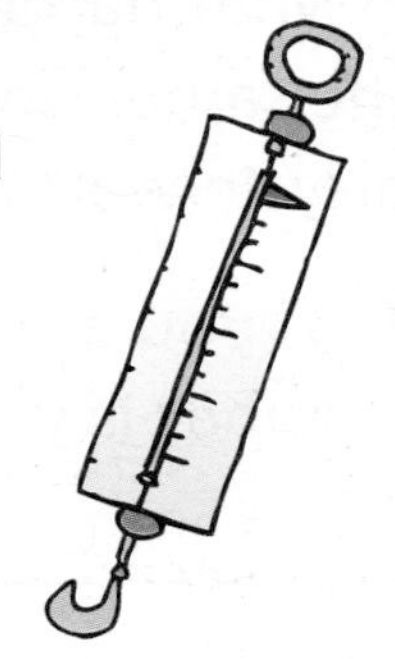

1. Hook a spring scale to a cardboard box.
2. Drag the box.
3. What happens when the scale shows a large reading?

____

____

4. What happens when it reads zero?

____

____

Quick Check

A cause makes something happen. An effect is what happens. Circle the cause of velocity being the same. Underline an effect of riding your bicycle west at a speed of 10 miles per hour.

1. If you are traveling north at 50 miles per hour, what is your velocity?

____

2. If your velocity is 35 miles per hour, east, and you turn south and slow down to 20 miles per hour, did your velocity change? ________. If it changed, what is your new velocity? If not, how could you change your velocity?

____

3. What two things do you need to know in order to find an object's velocity?

____

4. Circle the object on this page that has the greater velocity.

Velocity

Velocity is the speed and the direction of an object. When the speed and direction of two objects are the same, their velocity is the same.

Suppose you are riding your bicycle west at a speed of 10 miles per hour. Your velocity is 10 miles per hour, west. A friend is riding south at the same speed. His velocity is 10 miles per hour, south. Your velocities are different because you are going in different directions.

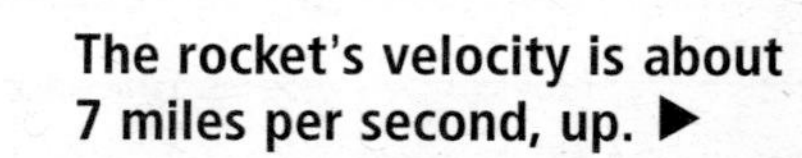

The rocket's velocity is about 7 miles per second, up. ▶

▲ The velocity of the two bikers is about 15 miles per hour, west.

Changing Velocity

Objects stop and start. They slow down and speed up. They turn. These are examples of acceleration. **Acceleration** is any change in the speed or the direction of an object in motion. So acceleration is any change of velocity. The greater the change in speed, the greater the acceleration.

The boat accelerates when it changes direction. ▶

▲ The skier accelerates when she changes speed and direction.

✓ Quick Check

1. What is acceleration?

____

____

2. What are some ways to accelerate?

____

____

3. Look at the table below. If the example of motion tells about acceleration happening, put **Y** in the box next to it. If it does not show acceleration, put an **N** in the box next to it.

Type of Motion	Is It Acceleration?
Someone skiing straight down a ski slope at the same speed	
A car stopping at a stop sign	
A person swimming around in circles at the same speed	
A car with a velocity of 60 km per hour, north	
A roller coaster beginning to go down a hill	

✓ Quick Check

Focus Skill A cause makes something happen. An effect is what happens. Circle the cause of an object changing speed or direction. Underline an effect that happens when tapping a ball lightly.

1. Complete the table below:

Cause	Effect
	The door moves away from you.
You pull a door.	
You tap a ball with your finger.	
	The ball takes off.

2. Why is it harder to stop a large truck than a small car when both have the same velocity?

____

____

____

Force and Acceleration

Pushes and pulls are **forces**. Forces can change an object's speed or direction. Forces can change an object's acceleration. The direction an object moves depends on the direction of the force. If you push a door, it moves away from you. That changes the door's acceleration.

The size of a force also has an effect. A greater force causes a greater acceleration. Tap a ball lightly with your finger. It may roll. Kick it as hard as you can. It takes off.

Mass and Acceleration

It is easier to push a bicycle than a car. That is because a car has more mass. It takes more force to change the speed or direction of an object with a large mass than one with a small mass. The reason mass has an effect is called *inertia*.

Inertia is a property of matter that keeps an object from changing speed or direction unless a force acts on it. A moving object tends to keep moving. An object at rest tends to stay at rest. The more mass an object has, the more inertia it has.

It is also harder to stop a car than a bicycle. This is because the car has greater *momentum*. Momentum is a property of motion that describes how hard it is to slow down or stop an object. Momentum depends on mass *and* velocity.

A car has more mass than a bicycle. So, it is harder to stop. Also, if the car bumps into something, it will cause more damage. Momentum increases if either mass or velocity increases.

Six dogs pull with more force than one dog.

Lesson Review

Complete these cause-and-effect statements.

1. When you know the speed and direction of an object, you can tell its ____________.
2. When a boat changes its speed, it ________________.
3. Both ______________ and ______________ depend on mass.
4. Fill in this graphic organizer.

A ________ is a push or pull	that can cause →	acceleration, a change of ____________.

Georgia Performance Standard

S4P3d Demonstrate the effect of gravitational force on the motion of an object.

Vocabulary Activity

In this lesson, you will learn about how gravity affects motion. Fill in the blanks with the correct vocabulary term.

1. ______________ is a force that pulls everything toward Earth.
2. The force that slows down or stops motion between objects that are touching is called ______________.
3. ______________ is the gravitational force that pulls on an object.
4. The force that acts between all objects and causes them to pull on each other is called ______________.

Student eBook www.hspscience.com

Lesson 3

How Does Gravity Affect Motion?

VOCABULARY
gravity
gravitation
weight
friction

Gravity is a force that pulls everything toward Earth. Gravity pulls sky divers to Earth.

Gravitation is a force that acts between all objects and causes them to pull on each other. Gravitation holds Earth in its orbit around the sun.

Weight is the gravitational force that pulls on an object. It is measured with a scale.

Friction is a force that slows down or stops motion between objects that are touching. A bike's brakes use friction.

Insta-Lab

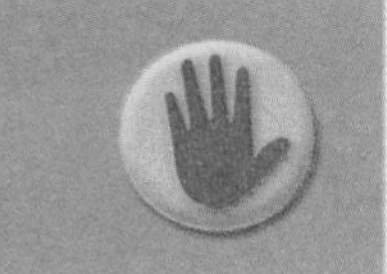

Get the Feel of Friction

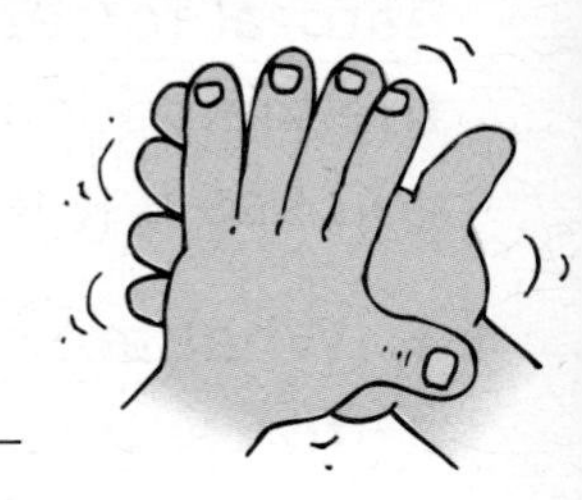

1. Rub your hands together. What happens?

____

2. Put a drop of cooking oil on your hands. Rub them together again. What happens?

____

3. Put on rubber gloves. Rub your hands together. What happened?

____

4. How is the friction different each time?

____

____

____

____

____

Quick Check

The main idea on these two pages is You use natural forces all the time. Details tell more about the main idea. Underline two details about how you use natural forces.

1. What are some ways in which we use magnetic forces and electricity?

____

____

____

____

2. Tell one way you used the force of your muscles today.

____

____

____

Natural Forces

You use natural forces all the time. When you walk, you use the force of your muscles. You use the forces of magnets and electricity in machines. You know about the force between electric charges. They hold together the tiny particles of all matter, including you. Even the sun and Earth pull on each other.

The sun pulls on Earth. Earth pulls on the sun with an equal and opposite force. ▼

Gravity

Gravity is a force that pulls you toward Earth. It is an example of gravitation. **Gravitation** is a force that acts between all objects. It causes them to pull on each other. Gravitation holds Earth in its orbit around the sun. It also helps hold the moon in its orbit around Earth.

▼ **Gravity pulls everything toward Earth.**

✓ Quick Check

1. How does gravity affect objects?

____

2. What would happen if gravity did not exist?

____

____

3. In the space below, draw two pictures. In the first picture, draw a baseball being thrown into the air. In the second, draw what has happened to the ball as a result of gravity.

Quick Check

Focus Skill The main idea on these two pages is Weight is the force of gravitation that pulls on an object. Details tell more about the main idea. Underline two details about the force of gravitation on an object.

1. How do we measure weight?

____

2. How is mass different from weight?

____

____

____

3. Which tool is used to measure mass?

____

Weight

Weight is the force of gravitation that pulls on an object. You measure this force in newtons (N) using a scale. One newton is about the weight of an apple.

Weight is not the same as mass. Mass is the amount of matter in an object. You measure mass on a balance. The unit of mass is the gram.

◀ **A scale measures weight.**

▼ **A balance measures mass.**

Do not confuse weight and mass. Your weight is the force of Earth's gravity pulling on you. If you were away from Earth, your weight would be different. On the moon, for example, your weight would be less. The moon's force of gravitation is less than Earth's. But your mass stays the same wherever you are.

The girl is being weighed on a doctor's scale. She weighs 250 newtons, or 56 pounds. ▼

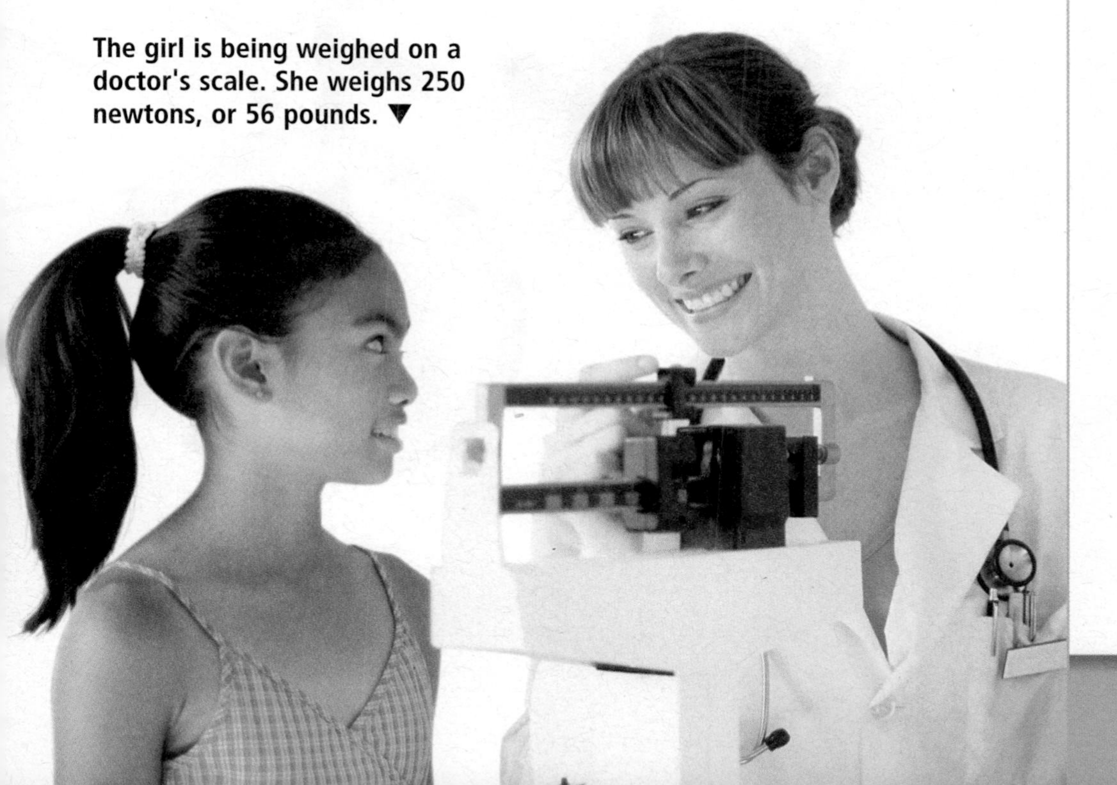

✓ Quick Check

1. How would your weight change on the moon?

____

Why?

____

____

2. How would your mass change on the moon?

____

Why?

____

____

____

✓ Quick Check

Focus Skill The main idea on these two pages is Friction is a force between two things that rub against each other. Details tell more about the main idea. Underline two details about friction.

1. What two things can friction do?

____

____

2. Write two examples of friction.

____

____

____

3. Why is friction a useful force?

____

____

____

Friction

Friction is a force between two things that rub against each other. Friction can slow down and stop the motion of things that touch each other. Friction also makes heat. Rub your hands together. Feel the heat.

Friction slows down and stops a bike.

You use friction all the time. When you stop your bike, the brakes press against the wheel and cause friction. The rubber soles of your sneakers increase friction, too. This keeps you from slipping.

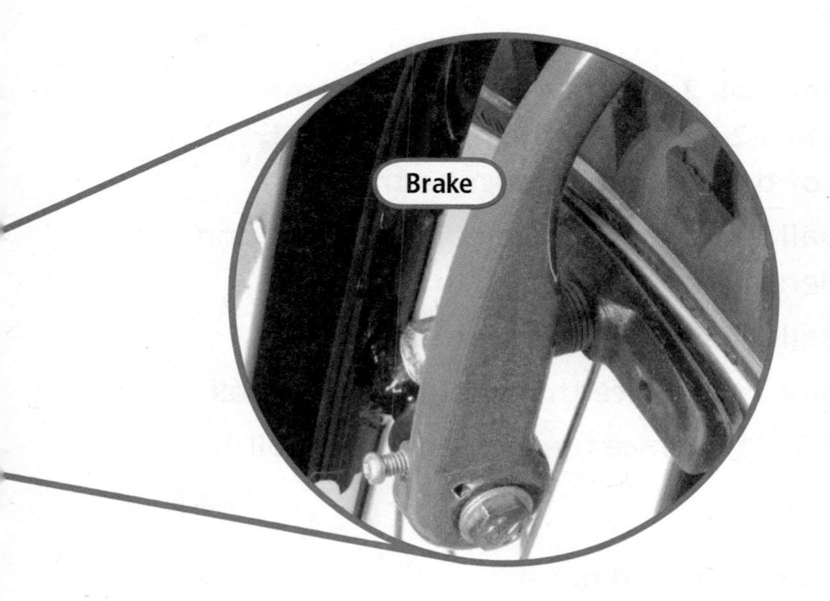

Lesson Review

Complete this main idea statement.

1. Gravity, gravitation, and friction are _________.

Complete these detail statements about forces.

2. When you use brakes to stop your bike, you are using ______________.
3. When you toss a ball into the air, ______________ pulls it back down to the ground.
4. Earth stays in its orbit because of ____________.
5. Fill in this graphic organizer.

Kinds of Forces

_______________	_______________	_______________
is a measure of the force of gravity.	is a force between all objects.	opposes motion between touching objects.

Fill in the circle in front of the letter of the best choice.

1. **You have four balls. Their masses are 100 grams, 200 grams, 300 grams, and 400 grams. Which one will accelerate MOST if they are all thrown with equal force?**

 ◯ A. the 100-gram ball
 ◯ B. the 200-gram ball
 ◯ C. the 300-gram ball
 ◯ D. the 400-gram ball

S4P3b

2. **A force is causing an object to move at a constant speed, as shown in the first picture. A second force causes the object to change direction, as shown in the second picture. If the two forces are equal in magnitude, how does the object's speed change?**

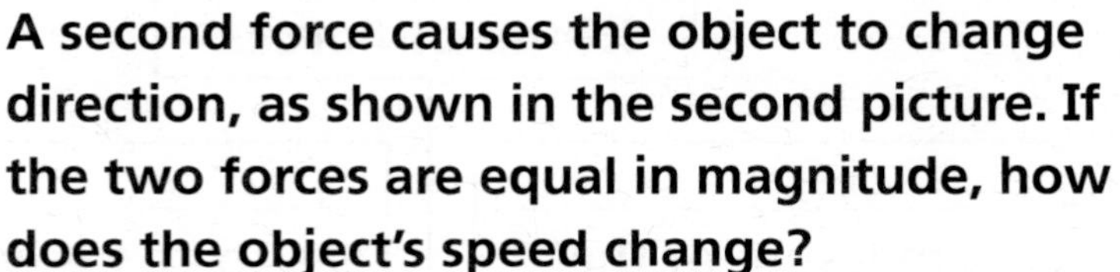

 ◯ A. Its speed increases.
 ◯ B. Its speed decreases.
 ◯ C. Its speed does not change.
 ◯ D. Its speed accelerates.

S4P3c

3. **The spheres shown are all made of pure iron. Which one needs the LEAST force to move?**

 ◯ A. ◯ B. ◯ C. ◯ D.

S4P3b

4. **Ball 1 and Ball 2 are traveling at the same speed. Ball 2 has twice the mass of Ball 1. Which of the following must be true?**

 ◯ A. Ball 1 and Ball 2 have equal volume and density.
 ◯ B. Ball 1 has twice the inertia of Ball 2.
 ◯ C. Ball 1 has twice the acceleration of Ball 2.
 ◯ D. Ball 2 has twice the momentum of Ball 1.

S4P3b

5. **If Earth's gravity did not act on the moon, the moon would**

 ◯ A. fly off into space in a straight line.
 ◯ B. crash into Earth.
 ◯ C. orbit the Earth faster.
 ◯ D. weigh more than it does.

S4P3d

6. A car is traveling at a speed of 100 kilometers per hour. Suddenly, it starts going faster. What caused the increase in speed?

- A. velocity
- B. a force
- C. inertia
- D. speed

S4P3c

Use the table below to answer question 7.

Mass of Ball (in grams)	Speed of Ball (in km per hour)
100	20
200	20
300	20
400	20

7. Which ball was thrown with the greatest force?

- A. the 100-gram ball
- B. the 200-gram ball
- C. the 300-gram ball
- D. the 400-gram ball

S4P3b

8. Tim is pushing a box across the floor. Mary decides to help him. What is the result?

- A. The mass of the box decreases.
- B. Friction is reduced by half.
- C. The box has less inertia.
- D. The box moves faster.

S4P3c

9. Which object requires the most force to push?

- A. car
- B. bus
- C. bicycle
- D. skateboard

S4P3b

10. What force keeps moons in orbit around planets?

- A. mass
- B. gravitation
- C. friction
- D. magnetism

S4P3d

11. An object is moving at a constant speed. The object changes direction and continues moving at the same speed. What caused the object to change direction?

- ◯ A. momentum
- ◯ B. a change in speed
- ◯ C. inertia and gravity
- ◯ D. a force acting on it

S4P3c

12. During which portion of its movement does the object have greater inertia?

- ◯ A. before it changes direction
- ◯ B. after it changes direction
- ◯ C. at the moment it changes direction
- ◯ D. The object's inertia does not change.

S4P3c

13. A truck has more inertia than a bicycle. Which characteristic of the truck accounts for this?

- ◯ A. its mass
- ◯ B. its volume
- ◯ C. its shape
- ◯ D.its age

S4P3b

14. A pitcher throws a baseball toward a batter. The batter hits the ball and sends it flying toward first base. What caused the ball to change direction?

- ◯ A. the acceleration of the pitcher
- ◯ B. the force of the batter
- ◯ C. the mass of the ball
- ◯ D. the inertia of the pitcher

S4P3c

15. Which of the following BEST explains why a ball that is hit into the air falls to the ground?

- ◯ A. acceleration
- ◯ B. density
- ◯ C. inertia
- ◯ D. gravity

S4P3d

16. The balls shown below travel toward you at equal speed. Which one requires the MOST force to stop?

◯ A. ◯ B. ◯ C. ◯ D.

S4P3b

17. A car is moving at a speed of 60 kilometers per hour. Its speed then increases to 70 kilometers per hour. What must be true of the car?

- ◯ A. Its direction of travel changed.
- ◯ B. Its mass increased.
- ◯ C. It accelerated.
- ◯ D. Its velocity decreased.

S4P3c

18. A car runs into a wall. What happens and why?

- ◯ A. The car stops because the wall exerts a force on it.
- ◯ B. The car bounces backward because the wall is elastic.
- ◯ C. The wall begins to travel at the same speed as the car did before it stopped.
- ◯ D. The friction between the car and the floor causes the car to stop.

S4P3c

19. A child throws two balls with equal force. One travels faster than the other. How do the balls compare?

- ◯ A. The faster one is heavier than the slower one.
- ◯ B. The faster one is lighter than the slower one.
- ◯ C. The lighter one has more mass than the slower one.
- ◯ D. The faster one has more mass than the heavier one.

S4P3b

20. Ball A travels 60 kilometers in 3 hours. Ball B travels 40 kilometers in 2 hours. Ball A is larger than Ball B. Which of the following is true?

- ◯ A. Ball A is traveling faster than Ball B.
- ◯ B. Ball B is traveling faster than Ball A.
- ◯ C. Ball A and Ball B are traveling at the same speed.
- ◯ D. The speed of the balls cannot be compared because the balls are not the same size.

S4P3b

Simple Machines

The six main kinds of simple machines make work easier by changing the force needed to move objects.

On this page, record what you learn as you read the chapter.

Essential Question

How do simple machines help people do work?

____

____

____

____

Essential Question

How do a pulley and a wheel-and-axle help people do work?

____

____

____

____

____

Essential Question

How do other simple machines help people do work?

____

____

____

____

____

____

Quick and Easy Project

Making an Elevator

Materials:
- pulley
- 2 paper cups
- string
- pennies
- tape
- box

Procedure

1. Tape the pulley to the underside of a box.
2. Pass the string over it.
3. Tape a paper cup to each end of the string.
4. Place pennies in one cup and then in the other. Watch your elevator move.

Draw Conclusions

1. How does the balance of weight affect the movement of your elevator?

____

____

2. Explain how the balance of weights relates to what you know about forces.

____

____

____

Georgia Performance Standard

S4P3a Identify simple machines and explain their uses (lever, pulley, wedge, inclined plane, screw, wheel and axle).

Vocabulary Activity

We use simple machines every day to help us do work. Use the vocabulary terms to fill in the blanks below to find out more about simple machines and work.

1. A ____________ ____________ has few or no moving parts.
2. The use of force to move an object from one place to another is ____________.
3. The point that a lever turns on is called the ____________.
4. A ____________ is a bar that pivots on a point that does not move.

Lesson 1

How Do Simple Machines Help People Do Work?

VOCABULARY
work
simple machine
lever
fulcrum

Work is the use of force to move an object from one place to another.

A **simple machine** has few or no moving parts. This wheelbarrow is a simple machine.

A **lever** is a bar that pivots, or turns, on a point that does not move. This fixed point is called the **fulcrum**. A hockey stick is one kind of lever.

Insta-Lab

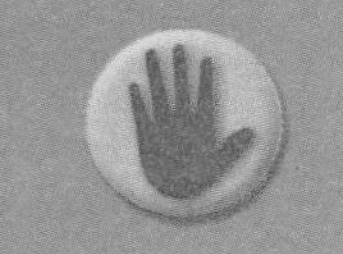

Lift It!

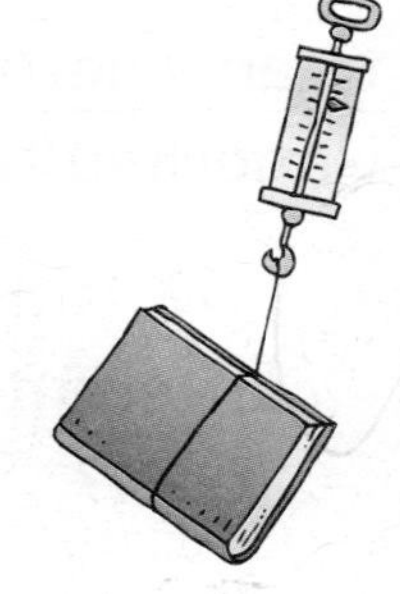

1. Tie a string around a book.
2. Hook a spring scale to the string.
3. How much force does it take to lift the book?

___

4. Move the string to one end of the book. Lift this end, leaving the other end on a table.
5. How much force is needed to lift the book now?

___

___

6. How is this action like using a wheelbarrow?

___

___

___

✓ Quick Check

The main idea of these two pages is Work is the use of force to move an object from one place to another. Details tell more about the main idea. Underline two details about work.

1. How is the scientific definition of *work* different from the everyday definitions you know?

   It is different cause [illegible] has more defintions of [illegible] force [illegible] other [illegible]

2. Is doing a math problem in your head work? Explain.

   No it is not cause it Does not show want your doing

Work and Simple Machines

We use the word *work* every day. But in science, this word has a special meaning. **Work** is the use of force to move an object from one place to another. The object must move in the same direction as the force used to move it. Lifting a box is work. But carrying a box is not work. That's because the box moves sideways with you as you carry it.

We use machines to help us do work. A **simple machine** is a machine with few or no moving parts. You apply just one force to make a simple machine work.

The wheelbarrow is a simple machine. The boy lifts the handles, and the pile of leaves goes up. He does the same amount of work as lifting the leaves by hand, but he uses less force. This simple machine makes work easier to do.

The boy uses a wheelbarrow to lift leaves.

✓ Quick Check

1. What is a simple machine?

2. In the picture of the wheelbarrow, circle the one moving part.

3. How does a simple machine such as a wheelbarrow change the way work is done?

✓ Quick Check

The main idea of these two pages is A lever is a bar that pivots, or turns, on a point that does not move. Details tell more about the main idea. Underline two details about types of levers.

1. Circle the fulcrum in the picture at the bottom of this page.
2. How does a pry bar help make work easier?
3. Tell why a pry bar is a lever.

Seesaw ▶

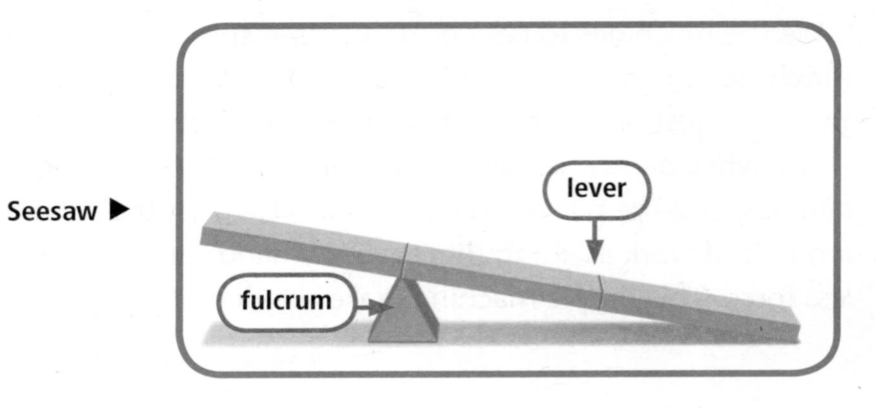

Levers

A **lever** is a bar that pivots, or turns, on a point that does not move. This fixed point is called the **fulcrum**. A seesaw is one kind of lever. The board of the seesaw is the bar. The point in the middle is the fulcrum. When you push down on one end, the other end goes up. A wheelbarrow and a broom are two other kinds of levers.

◀ **Find the lever and fulcrum in this hand truck.**

A pry bar is used to lift nails. ▼

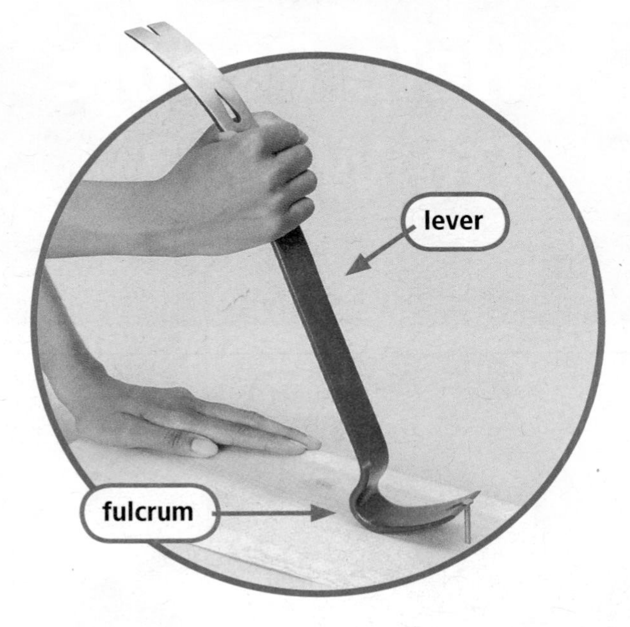

A pry bar is a lever, too. You can use it to pry a nail out of a board. You pull down on one end. The nail at the other end moves up. The pry bar makes work easier. After all, you couldn't pry the nail loose with just your hands!

Lesson Review

Complete the main idea statement.

1. A Simple Machine has few or no moving parts and needs just one force to make it work.

2. Fill in this graphic organizer.

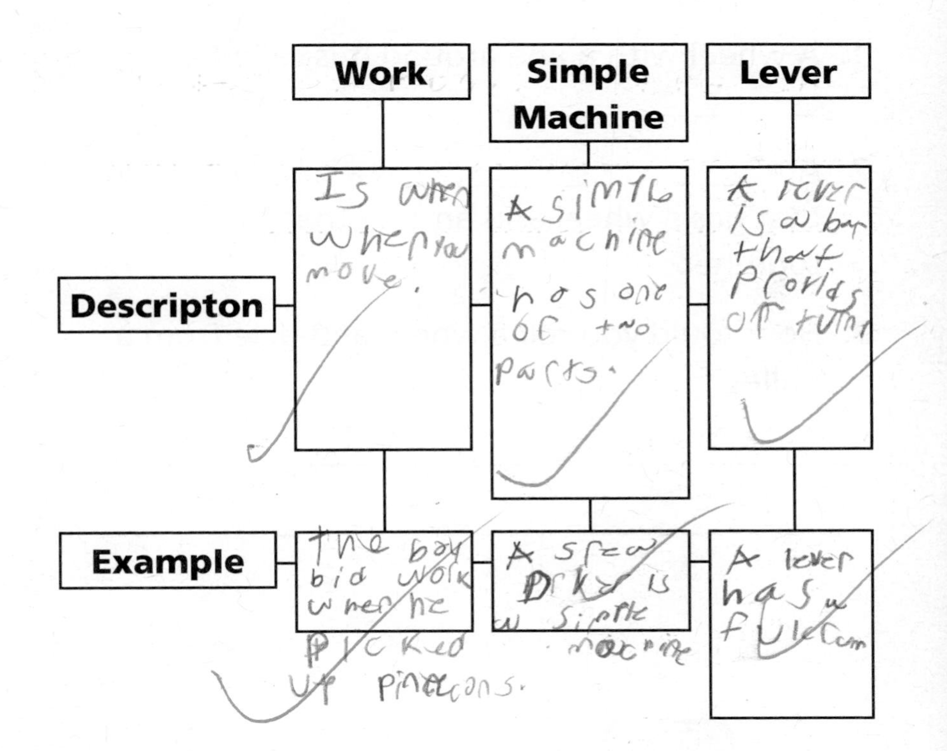

Georgia Performance Standard

S4P3a Identify simple machines and explain their uses (lever, pulley, wedge, inclined plane, screw, wheel and axle).

Vocabulary Activity

In this lesson, you will learn about more simple machines. Fill in each blank below with the correct vocabulary term.

1. A wheel with a line around it is called a __________.

2. A ______________________ is a simple machine that has a wheel and an axle that turn together.

3. How could you tell a wheel-and-axle from a pulley?

____

____

Go online ▸ Student eBook www.hspscience.com

Lesson 2

VOCABULARY
pulley
wheel-and-axle

How Do a Pulley and a Wheel-and-Axle Help People Do Work?

A **wheel-and-axle** is a simple machine that has a wheel and an axle that turn together. A doorknob is an example of a wheel-and-axle.

A **pulley** is a wheel with a line around it. This simple machine makes it easier to lift or move things.

Insta-Lab

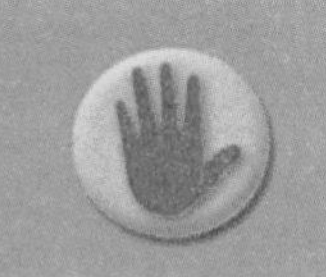

A Model Wheel-and-Axle

1. You can use a 2-L bottle to model a wheel-and-axle.
2. Hold a 2-L bottle by the cap.
3. Have a partner turn the bottle by the base so it tightens the cap.
4. Can you stop the bottle from turning? ________

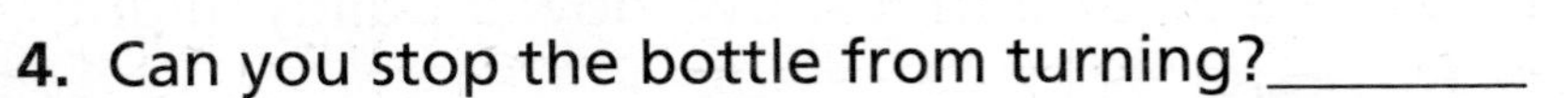

5. Roll the bottle along your desk by turning the cap.
6. How far do your fingers move?

   ____

7. How far does the bottle roll?

   ____

   ____

8. What part of the model represented the wheel?

   ______________ The axle? ______________

Quick Check

The main idea of these two pages is Like all simple machines, a pulley changes the way work is done. Details tell more about the main idea. Underline two details about how pulleys help do work.

1. List two types of work that pulleys can do.

____

2. Draw a picture that shows how a pulley might be used.

Pulleys

A **pulley** is a wheel with a line wrapped around it. The line may be a cord, a rope, or a chain. Small pulleys are used to raise shades. Large pulleys are used to lift machines and other heavy objects.

Like all simple machines, a pulley changes the way work is done. It changes the direction of the force. If you pull down on one end of the line, the other end goes up.

A single pulley changes only the direction of the force. Adding pulleys can lower the force needed to do work. You use less force, but you have to pull a greater distance. You use groups of pulleys to lift heavy loads. For example, a mechanic can use a large group of pulleys to lift an engine out of a car.

This man is using a large group of pulleys to increase his lifting force.

✓ Quick Check

1. In what two ways does using more than one pulley change the work that is done?

____

____

2. Why would a person want to use groups of pulleys?

____

3. Who might use a group of pulleys?

____

____

4. Tell how a single pulley changes the way work is done.

____

____

✓ Quick Check

Focus Skill The main idea of these two pages is A wheel-and-axle is a simple machine made of a wheel and an axle joined together. Details tell more about the main idea. Underline two details about a wheel-and-axle.

1. How does a wheel-and-axle used in a faucet help make work easier?

____

____

2. Name two examples of a wheel-and-axle.

____

3. Tell what a wheel-and-axle must do to be a simple machine.

____

____

Wheel-and-Axles

A **wheel-and-axle** is made of a wheel and an axle joined together. An axle is the bar on which the wheel turns. To be a simple machine, these parts must turn together. When you turn the wheel, the axle turns with it.

A faucet is a wheel-and-axle. When you turn the handle, the axle turns. But you use less force than if you turned just the axle. This makes work easier.

A faucet has a wheel and an axle that turn together. ▼

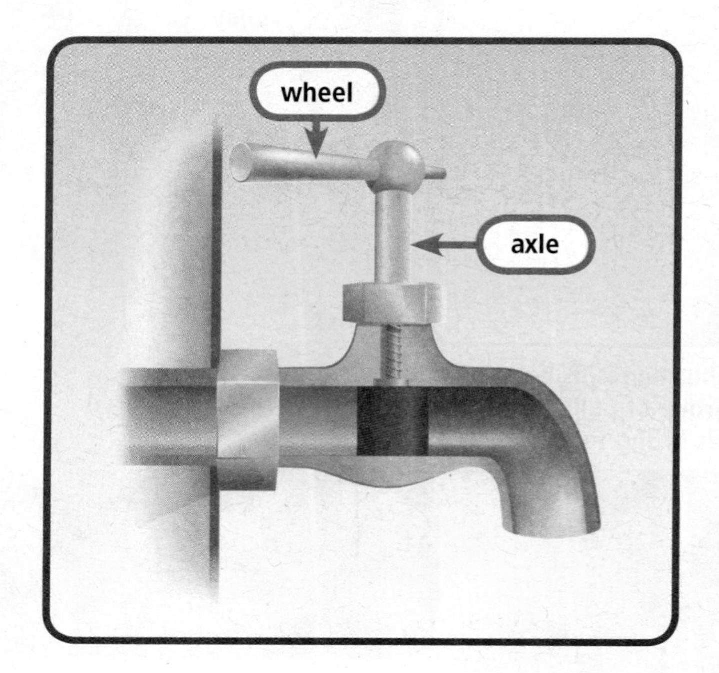

A salad spinner is another wheel-and-axle. The basket is the wheel. The crank is the axle. When you turn the axle, the basket spins. You have to move the axle just a short distance to move the outside edge of the basket a greater distance. You use more force to turn the crank, but you don't have to move it very far.

A salad spinner is a wheel-and-axle. ▼

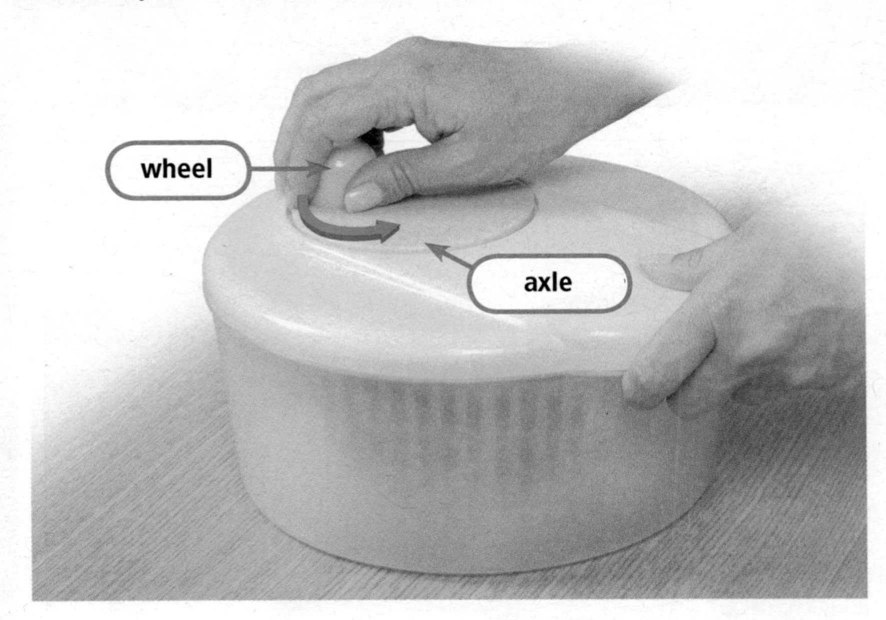

Lesson Review

Complete the main idea statement.

1. A pulley and a wheel-and-axle are both ___________ ___________.

Complete these detail statements about simple machines.

2. A single ______________ makes work easier by changing the direction of the force you use.
3. The two parts of a wheel-and-axle are joined so that they always move ______________.
4. In a faucet, the ______________ is the wheel.
5. Fill in this graphic organizer.

A pulley	A wheel-and-axle
is made up of	
with	

Georgia Performance Standard

S4P3a Identify simple machines and explain their uses (lever, pulley, wedge, inclined plane, screw, wheel and axle).

Vocabulary Activity

In this lesson, you will learn about other simple machines. Fill in the table below with the vocabulary terms.

Vocabulary Term	Definition
	a post with an inclined plane wrapped around it
	two inclined planes put back-to-back
	a slanted surface

Student eBook
www.hspscience.com

How Do Other Simple Machines Help People Do Work?

VOCABULARY
inclined plane
screw
wedge

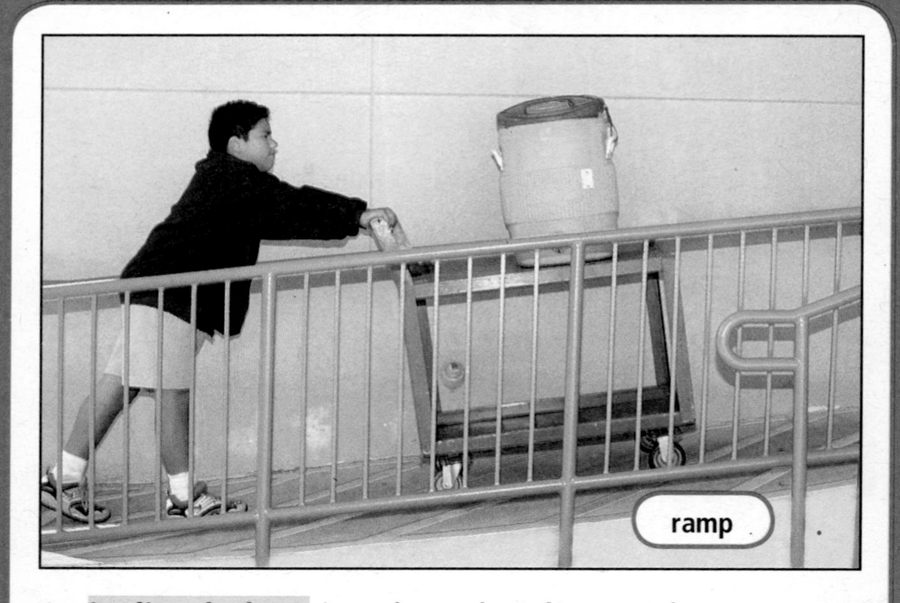

An **inclined plane** is a slanted surface, such as a ramp.

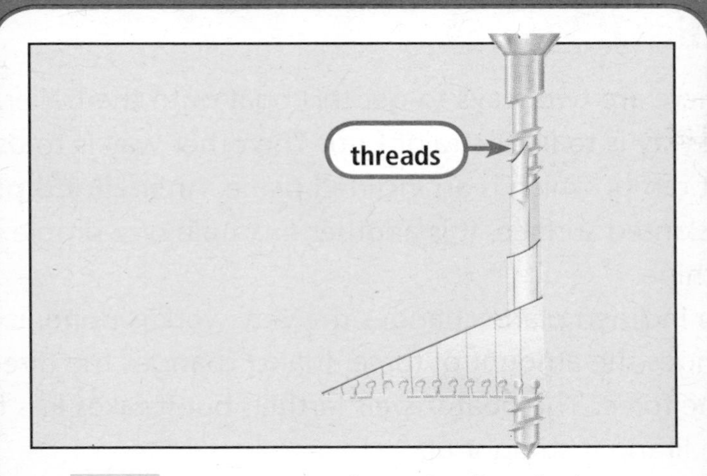

A **screw** is a post with an inclined plane wrapped around it.

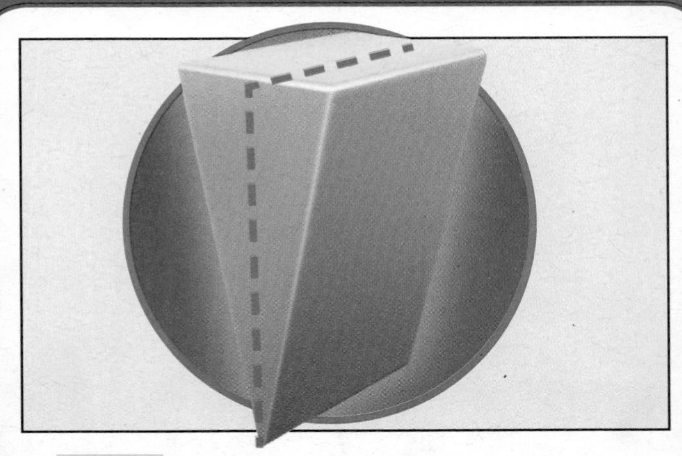

A **wedge** is two inclined planes put back-to-back. A wedge is used to split or cut things.

Insta-Lab

Spreading Spines

1. Work with a partner. One person should press two books tightly together.
2. The other person should insert the narrow edge of a wedge-shaped building block or doorstop between the books.
3. That person should then press gently toward the books.
4. What happens to the books?

____

5. Change roles, and do this again.

Quick Check

The main idea on these two pages is An inclined plane changes the way work is done. Details tell more about the main idea. Underline two details about how inclined planes change the way work is done.

1. What are two ways to get the boat in the picture onto the trailer?

____

2. How does an inclined plane change the way work is done?

____

____

Inclined Planes

There are two ways to get this boat onto the trailer. One way is to lift it straight up. The other way is to use a boat ramp, which is an inclined plane. An **inclined plane** is a slanted surface. It is another example of a simple machine.

An inclined plane changes the way work is done. It changes the amount of force. It also changes the direction of the force. The boat travels farther, but it takes less force than lifting it straight up.

▼ **A boat ramp is an inclined plane.**

A hill is an inclined plane. Look at these bikers. Both ride up inclined planes to get to the top. But one bike path is much steeper than the other. The steeper path is shorter. But the rider must pedal harder. The rider on the less steep path uses less force. But that rider must travel farther.

▼ **Which inclined plane is longer? Which takes more force to climb?**

✓ Quick Check

1. What is a detail about this main idea: *A hill is an inclined plane*?

____

____

2. When an inclined plane is made less steep, less force is needed, but what else changes? How does it change?

____

3. Tell how an inclined plane changes the way work is done.

____

____

____

____

4. Circle the person who is using more force.

✓ Quick Check

Focus Skill The main idea on these two pages is Wedges and screws are inclined planes. Details tell more about the main idea. Underline two details about wedges and screws.

1. How does a screw use an inclined plane?

____

2. List two examples of screws.

____

3. How does a wedge change the direction of force?

____

____

____

4. Tell how a wedge is like an inclined plane.

____

____

Screws

A **screw** is a post with threads wrapped around it. These threads are an inclined plane that curls around the post.

A drill bit is an example of a screw. By turning the bit around, you use less force than pushing the bit straight into the wood.

A nut and a bolt are also examples of screws. Their threads slide along each other. The nut and bolt keep two pieces of wood together.

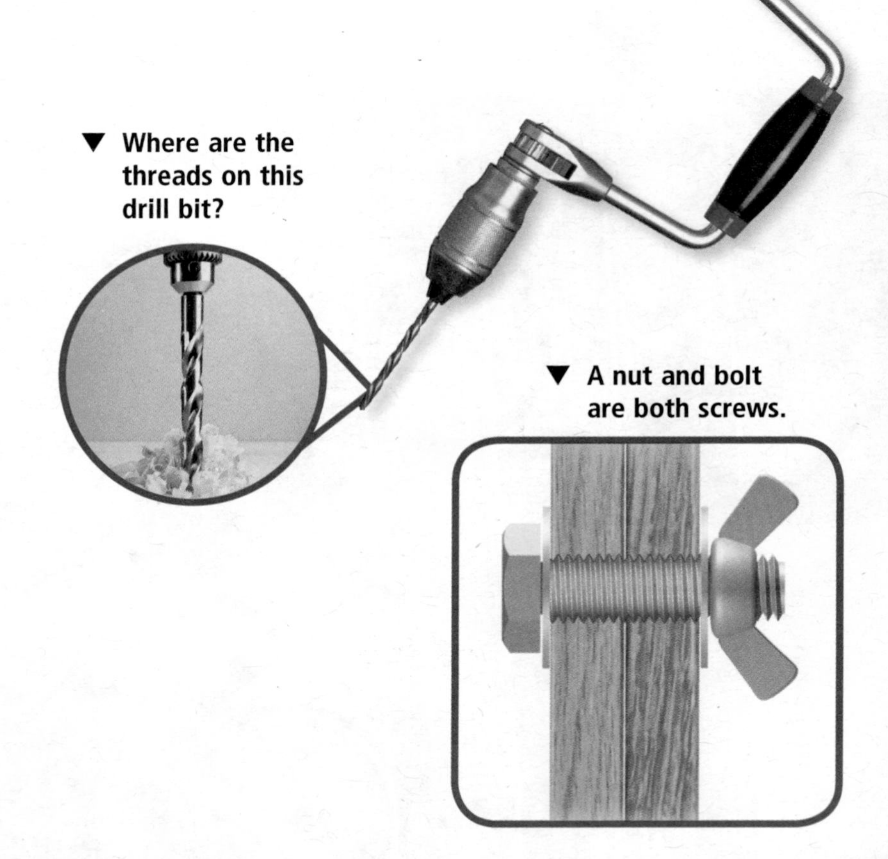

▼ Where are the threads on this drill bit?

▼ A nut and bolt are both screws.

Wedges

Like screws, wedges also use inclined planes. A **wedge** is two inclined planes put back-to-back.

You know that an inclined plane changes the direction of a force. A wedge does this, too. When you push down on a knife to cut an onion, the two parts of the onion move sideways, away from each other.

▲ **A knife is a wedge.**

Lesson Review

Complete this main idea statement.

1. The inclined plane, screw, and wedge are all ____________ ____________.

Complete these detail statements about simple machines.

2. Every inclined plane is a ____________ surface.
3. The threads of a screw are really an ________ ________ wrapped around a post.
4. A ____________ is two inclined planes put back-to-back.
5. Fill in this graphic organizer.

<table>
<tr><th></th><th colspan="2">Inclined Plane</th><th colspan="2">Screw</th><th>Wedge</th></tr>
<tr><td>Main Idea</td><td colspan="2"></td><td colspan="2"></td><td></td></tr>
<tr><td>Details</td><td></td><td></td><td></td><td></td><td></td></tr>
</table>

Fill in the circle in front of the letter of the best choice.

1. **Which of these simple machines is also an inclined plane?**
   - ◯ A. pulley
   - ◯ B. lever
   - ◯ C. screw
   - ◯ D. wheel-and-axle

   S4P3a

2. **How many forces must be applied to make a simple machine work?**
   - ◯ A. one
   - ◯ B. two
   - ◯ C. three
   - ◯ D. four

   S4P3a

3. **Which two simple machines are MOST alike?**
   - ◯ A. screw and lever
   - ◯ B. lever and wedge
   - ◯ C. pulley and wheel-and-axle
   - ◯ D. screw and inclined plane

   S4P3a

4. **Which of the following would be the BEST simple machine to use to slice celery?**
   - ◯ A. wedge
   - ◯ B. pulley
   - ◯ C. screw
   - ◯ D. inclined plane

   S4P3a

5. **Which of these simple machines will NOT work without a fulcrum?**
   - ◯ A. lever
   - ◯ B. wheel-and-axle
   - ◯ C. wedge
   - ◯ D. single pulley

   S4P3a

6. **Which of the following is NOT an example of a screw?**
   - ◯ A. nut
   - ◯ B. nail
   - ◯ C. bolt
   - ◯ D. drill bit

   S4P3a

7. **Which simple machine allows you to lift an object by using less force, although you must move the object a greater distance?**

◯ A. inclined plane

◯ B. single pulley

◯ C. wedge

◯ D. wheel-and-axle

S4P3a

8. **Which of the following is NOT a lever?**

◯ A. a baseball bat

◯ B. a wheelbarrow

◯ C. a broom

◯ D. a ramp

S4P3a

9. **Fran uses a rake to clear away the leaves in her yard. What simple machine does she use?**

◯ A. lever

◯ B. wedge

◯ C. inclined plane

◯ D. wheel-and-axle

S4P3a

Use the diagram below to answer question 10.

Output force

Input force

10. **Joey uses a tool to pry the top off a paint can. What type of simple machine is he using?**

◯ A. wedge

◯ B. screw

◯ C. lever

◯ D. inclined plane

S4P3a

11. **Which of these simple machines is LEAST like the other three?**

◯ A. screw

◯ B. inclined plane

◯ C. wedge

◯ D. pulley

S4P3a

12. Which is an example of a wedge?

- ◯ A. a bottle opener
- ◯ B. a doorstop
- ◯ C. an elevator
- ◯ D. a wheel-and-axle

S4P3a

13. Which of the following is a lever?

- ◯ A. ceramic cup
- ◯ B. measuring tape
- ◯ C. snow shovel
- ◯ D. ballpoint pen

S4P3a

14. To use a pulley to change both the amount and direction of the applied force, you must

- ◯ A. combine it with a lever.
- ◯ B. hook it up with another pulley.
- ◯ C. place it on an inclined plane.
- ◯ D. attach it to the wall with screws.

S4P3a

15. Which kinds of simple machines make up scissors?

- ◯ A. wheel-and-axle and pulley
- ◯ B. pulley and inclined plane
- ◯ C. lever and wedge
- ◯ D. wedge and inclined plane

S4P3a

16. Which of these is NOT a simple machine?

- ◯ A. broom
- ◯ B. pry bar
- ◯ C. rake
- ◯ D. lawn mower

S4P3a

17. **Which of the following tools contains two levers?**

○ A. ○ B. ○ C. ○ D.

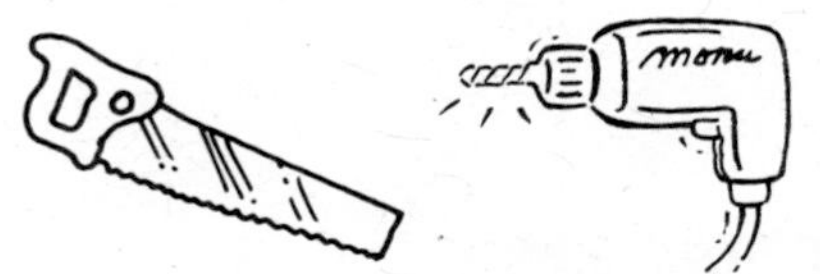

S4P3a

18. **Which simple machine could you use to hold two pieces of wood together?**

○ A. lever
○ B. pulley
○ C. screw
○ D. wheel-and-axle

S4P3a

19. **In science, which of the following is an example of work?**

○ A. going to a job
○ B. pushing against the floor
○ C. reading a book
○ D. lifting a chair off the floor

S4P3a

20. **Which two simple machines are inclined planes?**

○ A. pulley and wheel-and-axle
○ B. lever and pulley
○ C. screw and wedge
○ D. screw and wheel-and-axle

S4P3a

Food Energy in Ecosystems

All living things need energy and matter to live and grow.

On this page, record what you learn as you read the chapter.

Essential Question

What is an ecosystem?

____

____

____

____

Essential Question

How does energy flow through an ecosystem?

____

____

____

____

Essential Question

What are the roles of producers, consumers, and decomposers?

____

____

____

____

Essential Question

What factors influence ecosystems?

____

____

____

____

Quick and Easy Project

Energy Pyramid

Procedure

1. Identify producers, herbivores, carnivores, and omnivores that live in your area. List them on scrap paper.
2. Use the ruler to draw a large pyramid on the white paper. Divide the pyramid into three or four levels, depending on the kinds of living things you have identified.
3. Arrange some or all of these living things on your energy pyramid. Draw only one animal at the top level, ten at the next level, and so on.

Materials:

- scrap paper
- ruler
- large sheet of white paper
- colored pencils

Draw Conclusions

1. Do all the things from your list fit into your pyramid?

   ____

   ____

   If not, why not?

   ____

   ____

   ____

2. If you lived in a different kind of habitat, how would your energy pyramid be different?

   ____

   ____

Georgia Performance Standard

S4L1 Students will describe the roles of organisms and the flow of energy within an ecosystem.

Vocabulary Activity

Ecosystems contain many different parts. Use the pictures to answer the questions about the vocabulary terms.

1. What parts in the picture for the term *environment* are parts of that area's environment?

____

____

2. What parts in the picture for the *term* ecosystem interact?

____

____

3. What kind of population is shown in the picture?

____

Student eBook
www.hspscience.com

Lesson 1

What Is an Ecosystem?

VOCABULARY
environment
ecosystem
population
community

It is important to protect the **environment**. The environment is made up of living and nonliving things.

Changing the direction of a flowing river can hurt the river's **ecosystem**.

People sometimes build homes where prairie dogs live. This makes the space for the animals smaller. Then the **population** of prairie dogs must live with this change.

If a whole population of trees dies, the **community** of birds and animals that depended on them will suffer and change.

Insta-Lab

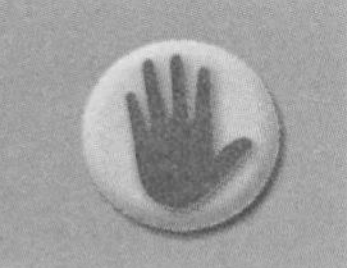

Eeek! Oh System!

1. Work with a partner to list some of the populations in your school ecosystem.
2. Compare your list with those of other students.
3. Did you list the same populations?

____

4. What populations did you both list?

____

____

____

Quick Check

The main idea of these two pages is All the things that interact in an environment make up an ecosystem. Details tell more about the main idea. Underline two details about ecosystems.

1. What are three nonliving things in the environment around you right now?

____

____

2. What are the two main parts of an environment?

____

Ecosystems

All the things around you make up the **environment**. An environment includes living things. Living things are people, animals, and plants. Nonliving things are in an environment, too. Nonliving things include water and air. Soil and weather are nonliving too. Your room environment has people, paper, light, air, and much more.

All the things that interact in an environment make up an **ecosystem**. An ecosystem has two parts—living things and nonliving things.

An ecosystem can be small or large, wet or dry, cold or hot.

The parts of an ecosystem work together. Dead plants make the soil healthy. Soil and water help new plants grow. Animals eat plants.

◀ **Under a rock, you might find an ecosystem of worms, stones, and bits of wood.**

An ecosystem can be under a rock. Wet soil, insects, and worms are in that ecosystem.

An ecosystem can be large. A forest is a large ecosystem. Trees, animals, and rocks are in a forest ecosystem.

An ecosystem has a *climate*. The climate can be hot, warm, or cold. The climate can be wet or dry. The desert has a dry climate. In a forest, the climate might be cool and damp. Climate change affects ecosytems.

In a forest ecosystem, the climate can be rainy or dry, cold or hot.

✓ Quick Check

1. What might you find in an ecosystem under a rock?

____

2. What kind of climates can forests have?

____

____

3. An ecosystem has living parts and nonliving parts. Fill in the table below with living and nonliving things in the pictures on this page.

Living	Nonliving

✓ Quick Check

Focus Skill The main idea of these two pages is A population is a group of the same kind of plant or animal living in the same ecosystem. Details tell more about the main idea. Underline two details about populations.

1. Put a square around the population on this page.
2. An individual deer can be part of what kind of population?

____

3. Is an alligator living in Georgia a part of the same population as an alligator living in Alabama? Explain your answer.

____

____

____

Individuals and Populations

One deer is an *individual*. A group of deer is called a **population**. A population is a group of the same kind of plant or animal living in the same ecosystem. A blue jay is an individual. A group of blue jays is a population. Robins are birds, too. But robins are a different population, because robins are not the same as blue jays.

The individual water lily is part of a population of water lilies.

◀ **The red-winged blackbird can live in different ecosystems.**

Some populations can live in different ecosystems. Some birds live in dry places. They can also live in wet places. If an ecosystem changes, the birds can fly somewhere else.

Some populations must live in only one kind of ecosystem. If the ecosystem changes, the population will have a hard time.

▲ **This ecosystem is a cypress swamp. This ecosystem is named for the main population of trees that live there.**

✓ Quick Check

1. How does a cypress swamp ecosystem get its name?

____

____

2. How is it helpful for populations to be able to live in different ecosystems?

____

____

3. Name a population you might see at a pond.

____

____

✓ Quick Check

The main idea of these two pages is A community is all the populations that live in the same place. Details tell more about the main idea. Underline two details about communities.

1. What kinds of populations live together in the Blue Ridge Mountains area community?

____

____

2. What is one way that the Blue Ridge Mountain area community is different from the Okefenokee Swamp community?

____

____

3. How can communities be different?

____

____

____

Communities

Plants, animals, and people live together. They live in a community. A **community** is all the populations that live in the same place.

The Blue Ridge Mountains area is a community. Many populations live in this community. There are different tree populations. There are different animal populations. There are insect and bird populations. All these populations live together in the Blue Ridge.

In the Blue Ridge, you will find many different populations.

There are different kinds of forests in Georgia. The Okefenokee Swamp has a forest community. The populations there are different from the Blue Ridge. For example, cypress trees grow in the swamp forest, while oak trees grow in the Blue Ridge forest.

Some populations, such as black bears, live in both communities.

▲ **These sand dunes on Cumberland Island are part of a coastal community.**

Lesson Review

Complete this main idea statement.

1. All the living and nonliving things in an area form an ______________.

Complete these detail statements about an environment.

2. An individual ladybug is part of a ladybug ______________.

3. Many different populations live in the same ______________.

4. Fill in this graphic organizer.

Main Idea: An ecosystem is made up of living and nonliving things.

Living Things	**Nonliving Things**
______________	______________
______________	______________
______________	______________
______________	______________

Georgia Performance Standard

S4L1a Identify the roles of producers, consumers, and decomposers in a community.

Vocabulary Activity

In this lesson, you will learn about the roles that living things play in an ecosystem. Use the definitions to fill in the table below.

	an animal that eats both plants and animals
	a living thing that makes its own food
	an animal that eats only plants
	an animal that eats other animals
	living things that break down parts of dead animals
	a living thing that must eat other living things to get energy

Student eBook
www.hspscience.com

Lesson 2

What Are the Roles of Producers, Consumers, and Decomposers?

VOCABULARY
producer
consumer
herbivore
carnivore
omnivore
decomposer

A plant is a **producer** because it can make its own food.

A ladybug is a **consumer** because it must eat other living things.

A cow is an **herbivore**. It eats only plants.

A bobcat is a **carnivore**. It eats other animals.

A bear eats both plants and animals. It is an **omnivore**.

Sow bugs eat dead plants. They are one kind of **decomposer**. Other decomposers break down parts of dead animals.

Insta-Lab

Who's an Omnivore?

1. Read the nutrition labels on several food containers.
2. Research the source of each food.
3. What does the food's source tell about the people who eat it?

____

____

____

____

✓ Quick Check

The main idea of these two pages is Producers make their own food, but consumers must eat other living things. Details tell more about the main idea. Underline two details about producers and consumers.

1. What do plants need from the sun? Why?

____

2. Why are plants called producers?

____

____

____

3. Why do you think plants would die without sunlight?

____

____

____

____

Producers and Consumers

Green plants make their own food. Their leaves use the energy in sunlight to make food. Any living thing that makes its own food is called a **producer**. Plants are producers. Producers can be small, like moss. Producers can be large, like an oak tree.

Plants use energy from sunlight to make food. Without the sun, plants would die.

A bobcat gets its energy from other consumers.

Some animals eat plants. The energy stored in the plants is used by the animal's body. Deer and cows eat plants.

Some animals eat other animals. Lions and hawks eat other animals. They do not eat plants.

An animal that eats plants or other animals is called a consumer. **Consumers** cannot make their own food the way plants can. Consumers must eat other living things. Consumers eat plants or other animals.

A horse gets energy from producers.

A black bear eats both plants and animals.

✓ Quick Check

1. Why do some animals eat plants or other animals?

____

____

2. Name two animals that eat plants.

____

3. Name two animals that eat other animals.

____

4. Circle the picture of the animal that eats both plants and animals.

5. Give one example of a producer and one example of a consumer.

____

____

____

Quick Check

The main idea of these two pages is There are three kinds of consumers. Details tell more about the main idea. Underline three details about kinds of consumers.

1. Give an example of an herbivore, a carnivore, and an omnivore.

2. Use the diagram at the bottom of the page to fill in the table below.

Type of Living Thing	What It Eats
	Nothing; It makes its own food from sunlight.
	producer
omnivore	
carnivore	

Kinds of Consumers

There are three kinds of consumers. A **herbivore** is an animal that eats only plants, or producers. Horses are herbivores. Squirrels and rabbits are herbivores. They also eat producers.

A **carnivore** is an animal. It eats only other animals. Lions are carnivores. A carnivore can be large, like a whale. A carnivore can be small, like a frog. Carnivores eat other consumers.

An **omnivore** is an animal. It eats both plants and animals. Bears are omnivores. So are many people. They eat producers and consumers.

This diagram shows how consumers get energy to live. The arrows show the direction of energy flow.

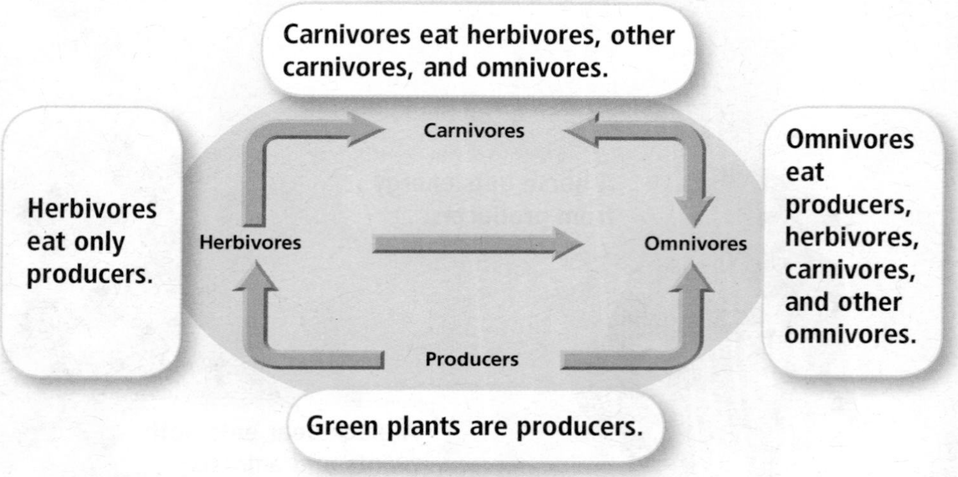

Carnivores eat herbivores, other carnivores, and omnivores.

Herbivores eat only producers.

Carnivores

Herbivores

Omnivores

Producers

Omnivores eat producers, herbivores, carnivores, and other omnivores.

Green plants are producers.

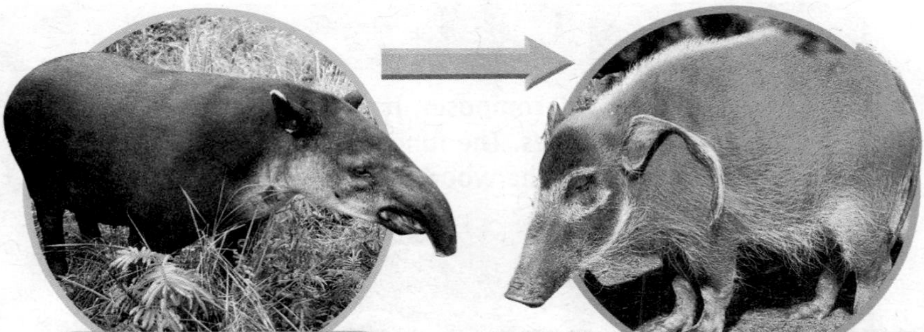

✓ Quick Check

1. Jaguars eat tapirs and river hogs. What kind of consumer are jaguars?

____

2. What kind of consumer is a tapir, and where does it get energy?

____

____

3. What kind of consumer is a river hog, and where does it get energy?

____

____

____

✓ Quick Check

The main idea of these two pages is A decomposer eats dead plants and animals. Details tell more about the main idea. Underline two details about decomposers.

1. Give two reasons that decomposers are important.

____

____

____

2. Name three types of decomposers.

____

____

Decomposers

A decomposer is a living thing. A **decomposer** eats dead plants and animals. Decomposers also break down waste. A decomposer can make soil healthy. Good soil helps plants grow. Animals eat the healthy plants. This keeps the animals healthy, too.

This fungus is a decomposer. It grows on dead trees. The fungus helps break down the wood.

Earthworms are decomposers. They eat dead plants.

Fungi are decomposers, too. Fungi break down wood from dead trees.

Many bacteria are decomposers. Some bacteria live in the soil. Some bacteria live in water. Bacteria break down parts of dead plants and animals.

Decomposers are very helpful. Without them, dead plants and animals would cover Earth.

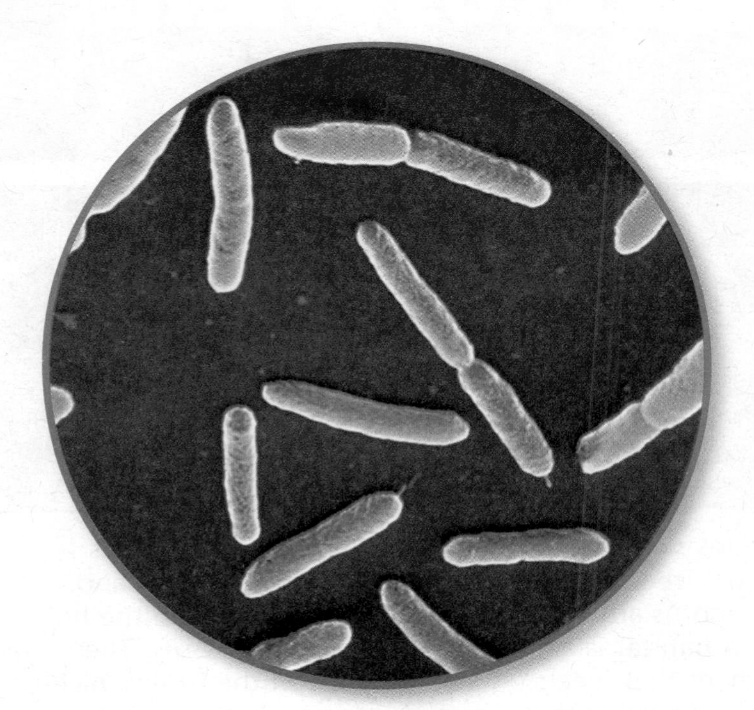

▲ **Bacteria can be found in water and soil.**

Lesson Review

Complete this main idea statement.

1. Herbivores, carnivores, and omnivores get energy from other ______________ things. ______________ make their own food.

2. Fill in this graphic organizer.

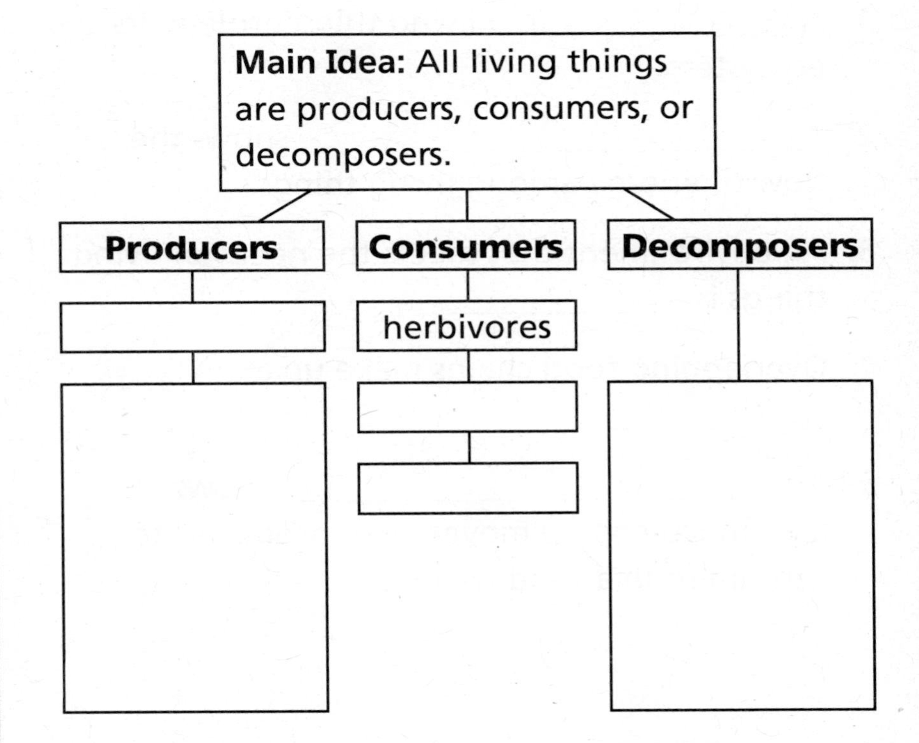

Georgia Performance Standard

S4L1b Demonstrate the flow of energy through a food web/food chain beginning with sunlight and including producers, consumers, and decomposers.

Vocabulary Activity

Use the vocabulary words to fill in the blanks.

1. A ______________ is a living thing's role in its ecosystem.
2. A ______________ ______________ shows the flow of energy among living things.
3. An environment that meets the needs of living things is a ______________.
4. Overlapping food chains make up a ______________ ______________.
5. An ______________ ______________ shows how much energy moves from producers to consumers in a food chain.

Student eBook
www.hspscience.com

Lesson 3

How Does Energy Flow Through an Ecosystem?

VOCABULARY
habitat
niche
food chain
prey
predator
food web
energy pyramid

A chipmunk's *habitat* is in the woods. Trees are all around. Acorns are on the ground. A **habitat** is an environment that meets the needs of a living thing.

Some kinds of turtles dig a hole in the sand. They lay eggs in the hole. They eat grasses. These are parts of the turtle's **niche**, or role, in its ecosystem.

Chipmunks are found in this owl's **food chain**, which shows the flow of energy among living things.

A hawk catches a fish. The fish is **prey**.

An owl is a **predator**. It eats mice and snakes.

Snakes eat mice. Hawks eat mice and snakes. These overlapping food chains make up a **food web**.

This **energy pyramid** shows how much energy moves from producers to consumers in a food chain.

Insta-Lab

Chain of Life

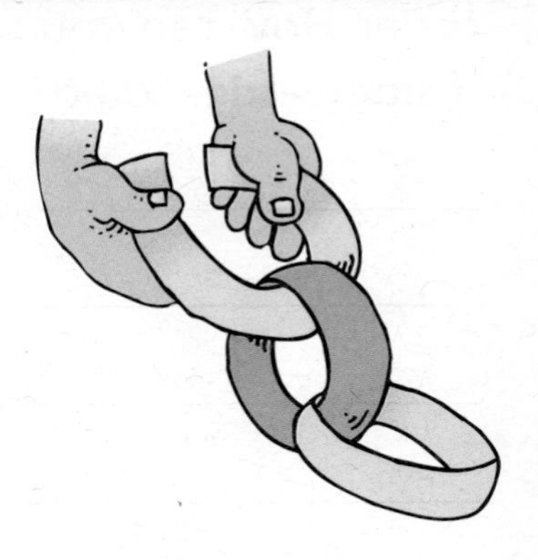

1. Cut white paper into strips that are 2.5 cm (1in.) by 12.5 cm (5 in.)
2. On each strip, write the name of a producer or a consumer.
3. Use glue or tape to combine the strips into paper food chains.
4. Give an example of a kind of food chain that ends with you.

____

____

✓ Quick Check

Focus Skill When you sequence things, you put them in order. How can you sequence what would happen if sidewinders died?

____

____

____

____

1. What are three types of organisms that might share a desert habitat?

____

2. Fill in the chart. Sequence forest, desert, and ocean habitats by how wet they are.

Amount of Water	Habitat
Dry	
Wetter	
Wettest	

Habitats

Plants and animals live in certain places. This is a habitat. A **habitat** is an environment that meets the needs of a living thing. Some plants and animals meet their needs in a desert. Some live in the sea. Others live where it is cold.

Different plants and animals can share a habitat. For example, spiders, snakes, and plants live together in the same desert habitat.

▲ Sidewinders, tarantulas, and sagebrush all live in the same habitat. The desert habitat meets all their needs.

This snake has a niche. Its niche helps to balance the number of small animals in its desert habitat.

Each living thing has a role, or **niche**. A niche is how a living thing works within the habitat. A niche includes these things:

- where a plant or animal lives.
- how it reproduces.
- where an animal gets food.
- how an animal stays safe.

A sidewinder is a snake. It has venom. The venom helps the snake kill mice and birds. Using venom to kill is part of the snake's niche.

If these snakes died, the desert would have too many mice and birds. There is not enough food in the desert for all these animals. They would begin to starve. The snake's niche keeps the number of small animals in balance.

✓ Quick Check

1. What things does a niche include?

____

____

____

2. How does the snake's niche help keep the number of small animals in balance?

____

____

____

3. Describe a sidewinder's niche.

____

____

____

✓ Quick Check

(Focus Skill) When you sequence things, you put them in order. How can you sequence a food chain?

____

____

____

1. What does a food chain show?

____

2. Why do you think all food chains begin with a producer?

____

____

____

Food Chains

Plants and animals depend on one another. Food energy moves from one living thing to another. A **food chain** shows the path of food energy. Most food chains start with producers, or plants that get energy from sunlight.

An oak tree is a producer. Acorns grow on oak trees. Chipmunks eat acorns. The chipmunk is a consumer. Then a hawk eats the chipmunk. The hawk is a consumer, too.

This is how food energy moves. It starts with the acorn. It moves to the chipmunk. Then it moves to the hawk. This is a food chain.

Consumers that are eaten are called **prey**. Consumers that eat prey are called **predators**. The chipmunks are prey. Hawks are predators. Owls are predators, too. Owls eat chipmunks. Predators, such as hawks and owls, compete for prey.

Acorns provide energy for the chipmunk.
Then the hawk eats the chipmunk.
The chipmunk provides energy for the hawk.

✓ Quick Check

1. What are animals that are eaten called?

____

2. What are consumers that eat other animals called?

____

3. Why do an owl and a hawk have to compete for prey?

____

____

____

4. Does a food chain start with a producer or a consumer?

____

5. Place an **X** on the organisim that gets its energy from sunlight.

✓ Quick Check

Focus Skill When you sequence things, you put them in order. How can you sequence a food web?

___

___

___

___

1. How are food chains and food webs related?

___

2. Look at the food web on this page. Which consumers eat grasshoppers?

___

3. Where can you find food webs?

___

Food Webs

A food chain shows how an animal gets energy. Different food chains can cross. For example, a hawk eats mice *and* small birds. When food chains cross, they make a **food web**. Food webs are on land. Food webs are in water, too.

Trace the different food chains, or paths, that make up this food web. For example, energy moves from plants to the mouse to the hawk. Energy also moves from plants to the snail to the sparrow to the hawk.

At the bottom of the food web are *first-level consumers*. Snails and insects are first-level consumers. Anything that eats a first-level consumer is called a *second-level consumer*. Snakes and birds are second-level consumers. They are eaten by the *top-level consumers*. Hawks and wolves are top-level consumers. Like all living things, when top-level consumers die, decomposers help break down their remains.

Quick Check

1. What are first-level consumers?

____

____

2. How do decomposers fit into a food web?

____

____

____

3. What happens after a first-level consumer eats a plant?

____

____

4. Use the picture on these two pages to fill in the sequence chart below.

Producer	First-level consumer	Second-level consumer	Top-level consumer

✓ Quick Check

Focus Skill When you sequence things, you put them in order. How can you sequence an energy pyramid?

____

____

____

____

____

____

____

____

1. What percent of energy is passed on to each level of the energy pyramid?

____

2. Circle first-level consumers. Place an **X** on a top-level consumer. Place a box around two second-level consumers.

Energy Pyramids

In a food chain, energy is passed on. Energy moves from one thing to another. An **energy pyramid** shows *how much* energy is passed on.

Producers are at the bottom of the energy pyramid. Plants get energy from the sun. Plants use about 90 percent of all this energy. They use the energy to grow. The extra 10 percent is stored. Plants store extra energy in their roots, stems, and leaves.

Then plants are eaten by squirrels and rabbits. These small consumers use up 90 percent of the energy from the plants. The other 10 percent of the energy is stored in their bodies.

The smaller animals are eaten by owls and foxes. These larger consumers use up 90 percent of the energy from the smaller animals. The other 10 percent is stored in their bodies.

Small amounts of energy are passed on. Energy moves from one living thing to the next. Each level passes a little energy to the next. The first-level consumers need many producers. That is why the bottom of the energy pyramid is wide.

This snake is a second-level consumer. ▼

Lesson Review

Complete these sequence statements.

1. A snake's ____________ helps to make a balanced habitat.
2. Every ____________ ____________ or ____________ ____________ starts with producers.
3. First-level consumers are eaten by ____________ ____________ ____________.
4. An energy ____________ ends with top-level consumers.
5. Fill in this graphic organizer.

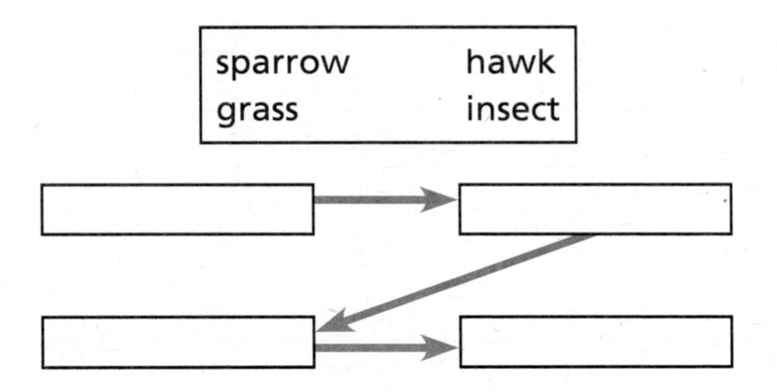

Georgia Performance Standards

S4L1c Predict how changes in the environment would affect a community (ecosystem) of organisms.

S4L1d Predict effects on a population if some of the plants or animals in the community are scarce or if there are too many.

Vocabulary Activity

Use the vocabulary words to fill in the left column. The first two are done for you.

Biotic or Abiotic?	Parts of an Ecosystem
abiotic	water
biotic	fish
	rocks
	trees
	horses
	air
	soil
	sand
	grass

Student eBook
www.hspscience.com

Lesson 4

What Factors Influence Ecosystems?

VOCABULARY
biotic
abiotic

Insects and plants need each other. They are **biotic**, or living, parts of the ecosystem.

Rocks and water are not alive. They are nonliving. They are the **abiotic** parts of the ecosystem.

Insta-Lab

Upsetting the Balance

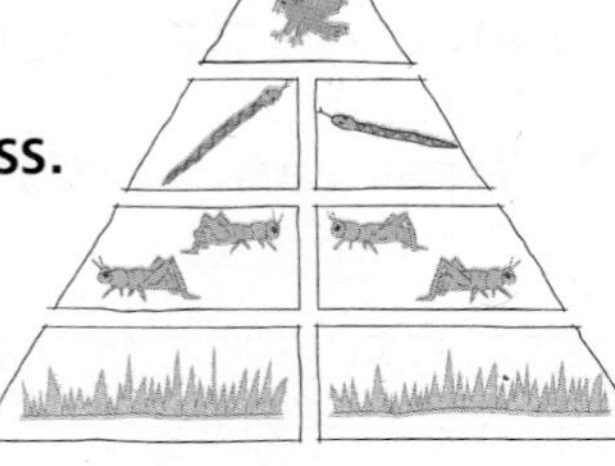

1. Draw an energy pyramid.
2. In the bottom row, draw grass.
3. In the next row, draw four grasshoppers.
4. In the next row, draw two snakes.
5. At the top, draw one hawk.
6. How would the hawk be affected if you took away half of the grass?

____

____

____

✓ Quick Check

Focus Skill A cause makes something happen. An effect is what happens. Circle a cause of ecosystems being changed. Underline an effect that a change in an ecosystem can cause.

1. How can plants affect animals?

____

____

____

2. How can animals affect plants?

____

____

Living Things Affect Ecosystems

Plants are the living parts of an ecosystem. Animals are living parts of an ecosystem, too. Living parts of an ecosystem are **biotic**. *Bio* means "life."

Plants affect animals in many ways. Plants are food for animals and insects. If a plant dies, there is less food. Plants, like trees, are also homes for animals and insects.

Animals affect plants in many ways. Birds and deer spread seeds. This causes plants to grow in new places. Animal droppings make the soil healthy. This helps the plants to grow. If too many animals eat a plant, they can kill it.

Plants and animals are biotic. They can change an ecosystem.

Tree leaves are a main source of food for deer. ▶

◀ Hungry moth caterpillars eat leaves on a tree.

A healthy tree is not hurt when a few caterpillars nibble on the leaves.

Too many caterpillars can eat all the leaves on a tree. This makes the tree weak.

Quick Check

1. Why are plants and animals biotic?

____

____

2. How can animals harm plants?

____

____

3. How can an insect change an ecosystem?

____

____

____

✓ Quick Check

Focus Skill A cause makes something happen. An effect is what happens. Circle a cause of ecosystems being changed. Underline an effect that an abiotic factor in the ecosystem causes.

1. What are abiotic factors?

____

2. Look at the picture on this page. What abiotic factors do you see?

____

Nonliving Things Affect Ecosystems

Plants and animals are living parts of an ecosystem. Ecosystems have nonliving parts, too. Sunlight, air, water, and soil are nonliving. These nonliving parts are **abiotic**. They are not alive. But they can change an ecosystem.

Look at the picture on these pages. The sun's light is shining. Plants need sunlight to grow and make food. Sunlight is an abiotic factor.

Water is abiotic. Plants need water to grow. Fish must live in water. Animals need to drink water. Water affects the ecosystem.

Soil is abiotic. Many plants need soil to grow. Some soil is dry and hard. Some soil is soft and wet. Soil affects the kinds of plants that can grow in an ecosystem.

✓ Quick Check

1. How does water affect an ecosystem?

____

____

2. How does soil affect an ecosystem?

____

____

____

3. What will happen to an ecosystem if a river dries up?

____

____

____

____

✓ Quick Check

Focus Skill A cause makes something happen. An effect is what happens. Circle a cause of change in ecosystems.

1. What are the parts of a climate?

____

____

____

2. Place an **X** on one desert ecosystem on the map.

3. How does a dry climate affect an ecosystem?

____

____

____

____

Climate Affects Ecosystems

What is the climate where you live? Is it hot, cold, or warm? Do you have a lot of rain, or is it dry where you live? Do you have a lot of sunlight? Climate is an abiotic factor. Climate is the amount of

- sunlight that shines in a place.
- rain that falls in a place.
- warm air that is in a place.
- cold air that is in a place.

Climate affects the soil in an ecosystem. Climate affects the plants that grow in an ecosystem. Climate also affects the animals that live in an ecosystem.

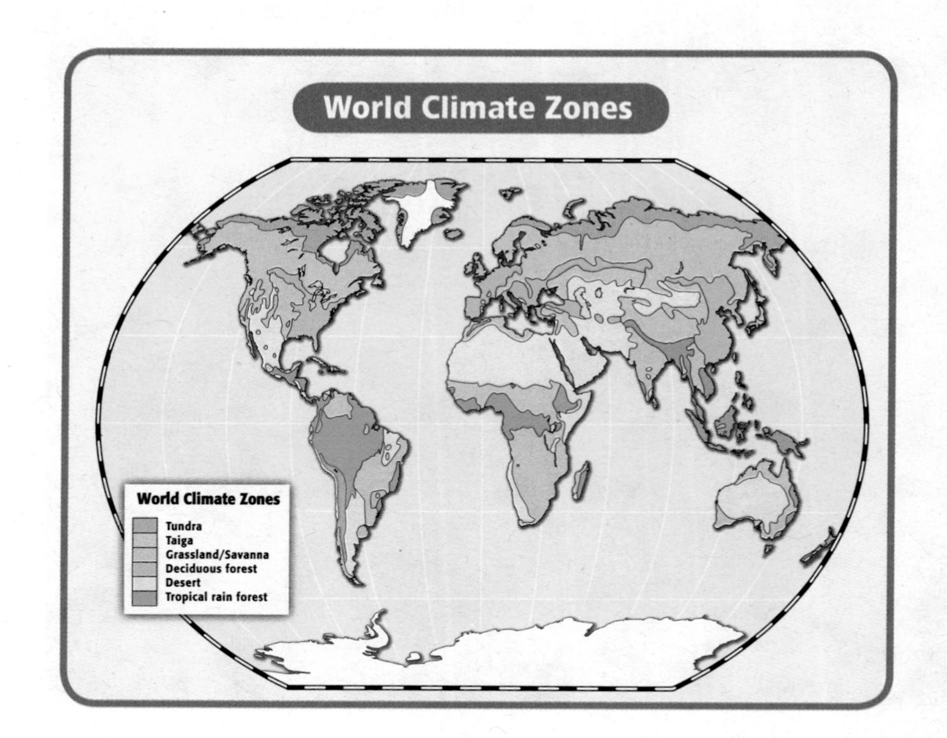

Some climates have four seasons.

A rain forest has a wet climate.

A desert has a dry climate.

Lesson Review

Complete the following cause-and-effect statement.

1. _______________ factors, such as birds spreading seeds, affect the ecosystem.

2. Fill in this graphic organizer.

Cause	Effect on Ecosystem
too many animals eat a plant	
river dries up	
increase in average temperature	

CRCT Practice

Fill in the circle in front of the letter of the best choice.

1. What is the role of a producer in a food web?

- ◯ A. to produce water
- ◯ B. to produce food
- ◯ C. to eat other organisms
- ◯ D. to decompose the bodies of dead animals

S4L1a

Use the diagram below to answer question 2.

2. According to this food web, what are two sources of energy for the bird?

- ◯ A. snail and grasshopper
- ◯ B. plant and fish
- ◯ C. rat and frog
- ◯ D. fish and frog

S4L1b

3. Decomposers are helpful in a food chain because

- ◯ A. they produce food from sunlight, water, and carbon dioxide gas.
- ◯ B. they eat other animals, keeping the populations of consumers down.
- ◯ C. they break down the remains of other organisms and add nutrients to the soil.
- ◯ D. they produce oxygen for other organisms to breathe.

S4L1a

4. What is the name for consumers that are eaten?

- ◯ A. carnivores
- ◯ B. fossils
- ◯ C. predators
- ◯ D. prey

S4L1a

5. Which abiotic factor is different after the fence is built?

- ◯ A. amount of light
- ◯ B. amount of water
- ◯ C. number of animals
- ◯ D. number of plants

S4L1c

Use the diagram below to answer questions 6 and 7.

6. Which is the BEST title for this diagram?

- ◯ A. A Wetland Food Chain
- ◯ B. A Wetland Food Web
- ◯ C. A Wetland Energy Pyramid
- ◯ D. A Wetland Population Pyramid

S4L1b

7. Which organism in the diagram is a first-level consumer?

- ◯ A. the plant
- ◯ B. the frog
- ◯ C. the grasshopper
- ◯ D. the bird

S4L1b

8. Which of the following animals is an herbivore?

- ◯ A. grizzly bear
- ◯ B. cow
- ◯ C. hawk
- ◯ D. lion

S4L1a

9. Fran observed rabbits, wildflowers, and a coyote. Which sequence correctly shows the transfer of energy among these things?

- ◯ A. coyote ➜ wildflowers ➜ rabbit
- ◯ B. coyote ➜rabbit ➜ wildflowers
- ◯ C. wildflowers ➜ coyote ➜ rabbit
- ◯ D. wildflowers ➜ rabbit ➜ coyote

S4L1b

10. What happens to a population if there are too many first-level consumers in a habitat?

- ◯ A. The number of producers in the habitat gets too low.
- ◯ B. There is not enough for second-level consumers to eat.
- ◯ C. It causes a change in the climate.
- ◯ D. Nothing happens to the habitat.

S4L1d

11. **Which organism is responsible for changing the sun's energy into food energy?**

◯ A. organism A

◯ B. organism B

◯ C. organism C

◯ D. organism D

S4L1a

12. **Which living things provide food for herbivores?**

◯ A. producers

◯ B. carnivores

◯ C. decomposers

◯ D. consumers

S4L1a

13. **What is the role of decomposers in an ecosystem?**

◯ A. to reduce predators in the area

◯ B. to provide food for the consumers

◯ C. to capture the sun's energy

◯ D. to break down dead plants

S4L1a

14. **Aphids are eating the rose-buds. A ladybug beetle is eating the aphids. Which term describes the aphids?**

◯ A. predator

◯ B. first-level consumer

◯ C. producer

◯ D. second-level consumer

S4L1b

15. **Picture an energy pyramid. Why does the number of living things decrease as you move toward the top?**

◯ A. The amount of sunlight decreases.

◯ B. There is not enough air in the environment for a large number of living things.

◯ C. The amount of energy available to living things decreases.

◯ D. There is not enough water in the environment for a large number of living things.

S4L1b

Use the drawing below to answer question 16.

16. How are these animals alike?

- ◯ A. Both are predators.
- ◯ B. Both are producers.
- ◯ C. Both are decomposers.
- ◯ D. Both are consumers.

S4L1a

Use the drawing below to answer question 17.

17. What is missing from the food chain?

- ◯ A. predator
- ◯ B. herbivore
- ◯ C. producer
- ◯ D. consumer

S4L1b

18. What would happen if the number of decomposers in an ecosystem was reduced?

- ◯ A. Dead plant and animal matter would break down more slowly.
- ◯ B. Temperatures would rise, causing dry weather.
- ◯ C. Other organisms would move away to find new sources of food.
- ◯ D. Rainfall would increase, causing flooding.

S4L1d

19. Which of the following describes a population?

- ◯ A. a family living together in a house
- ◯ B. a blue jay living alone in a tree
- ◯ C. all the trout living in a lake
- ◯ D. all the plants and animals in a forest

S4L1d

20. Bears are omnivores. Which meal would they eat?

- ◯ A. grass only
- ◯ B. berries and fish
- ◯ C. fish only
- ◯ D. herbivores only

S4L1b

Adaptations for Survival

Certain body parts and behaviors can help living things survive, grow, and reproduce.

On this page, record what you learn as you read the chapter.

Essential Question

What are physical adaptations?

____

____

____

____

____

____

Essential Question

What are behavioral adaptations?

____

____

____

____

____

____

Quick and Easy Project

Desert Leaves

Materials:
- 3 paper towels
- bowl with water
- large cookie sheet
- wax paper
- paper clips

Procedure

1. Wet each paper towel. Squeeze out some of the water so that it's not dripping.
2. Place one towel flat on the cookie sheet.
3. Roll up the second towel, and place it next to the flat towel.
4. Roll up the third towel. Roll the wax paper around it, and use a paper clip to keep it in place. Place this roll next to the other roll.
5. Place the cookie sheet in a sunny place. Wait 24 hours.
6. Unroll the towels. Feel each towel to see if it is still damp.

Draw Conclusions

1. Which towel kept the most water?

____

2. Which kept the least?

____

____

3. Why?

____

____

____

4. Use your results to explain how the structure of leaves of desert plants helps them conserve the desert's limited water.

____

____

____

Georgia Performance Standard

S4L2a Identify external features of organisms that allow them to survive or reproduce better than organisms that do not have these features (for example: camouflage, use of hibernation, protection, etc.).

Vocabulary Activity

Physical adaptations help living things survive in their environment. Use the pictures to help you answer the questions about the vocabulary terms.

1. What basic needs is the tiger meeting in the picture? ______________________

2. What adaptation does the alligator have for survival?

____

____

3. Why might it be hard for a bird to see the leaf moth?

____

____

Student eBook
www.hspscience.com

Lesson 1

What Are Physical Adaptations?

VOCABULARY
basic needs
adaptation
camouflage

Basic needs are the things that living things need to survive. Basic needs include food, water, air, and shelter.

An **adaptation** is a body part or a behavior that helps a living thing survive. Sharp teeth are an adaptation that helps an alligator catch its food.

Camouflage is a color or shape that helps an animal hide. Many birds do not notice this leaf moth because of its camouflage.

Insta-Lab

All Thumbs

1. Use masking tape to tape your partner's thumb to his or her hand.
2. Ask your partner to write, pick up a pencil, eat, and so on.
3. How is a thumb an adaptation?

____

____

____

Quick Check

The main idea of these two pages is All living things have the same basic needs. Details tell more about the main idea. Underline two details about the basic needs of animals.

1. How is the way that plants meet their need for food different from the way animals meet their need for food?

____

____

2. Look at the picture of the heron. What adaptation does it have to help it meet its need for food?

____

____

____

Basic Needs

All living things have the same basic needs. **Basic needs** are things a living thing needs to live and grow. These basic needs are food, water, shelter, and air.

Living things get food in different ways. Plants make their food. Animals get food by eating plants or other animals. Humans also get food by eating plants and animals.

A heron catches food. ▶

◀ A tiger drinks water.

Beavers build a shelter. ▼

Plants get water from rain. Most animals drink water. Some get it by eating foods with water in them.

Shelter takes many forms. Plants may grow in protected spots. Animals find shelter in different places. Some build or dig their own.

✓ Quick Check

1. Where do most plants get the water they need?

____

2. What are two ways in which animals can meet their need for shelter?

____

____

3. Look at the pictures on this page. How did the beaver meet its basic need for shelter?

____

4. In the table below, tell how you meet each of your basic needs.

Basic Need	How You Meet That Need
food	
water	
shelter	
air	

✓ Quick Check

The main idea of these two pages is To meet their needs, plants and animals have adaptations. Details tell more about the main idea. Underline two details about adaptations.

1. Look at the pictures on this page. How do you think the different type of mouth parts that are shown might be used?

____

____

____

2. How does camouflage help animals?

____

3. Name three examples of adaptations and how they are useful.

____

____

____

Adaptations

To meet their needs, plants and animals have adaptations. An **adaptation** is a body part or behavior that helps a living thing survive.

Animals have many adaptations. One is body covering. Different kinds of feet are also adaptations. They help animals move.

Some animals have long tongues or legs. Sharp beaks, teeth, or claws can help animals catch or eat food.

Another animal adaptation is called camouflage. **Camouflage** (KAM•uh•flahzh) is a color or shape that helps an animal blend with its environment.

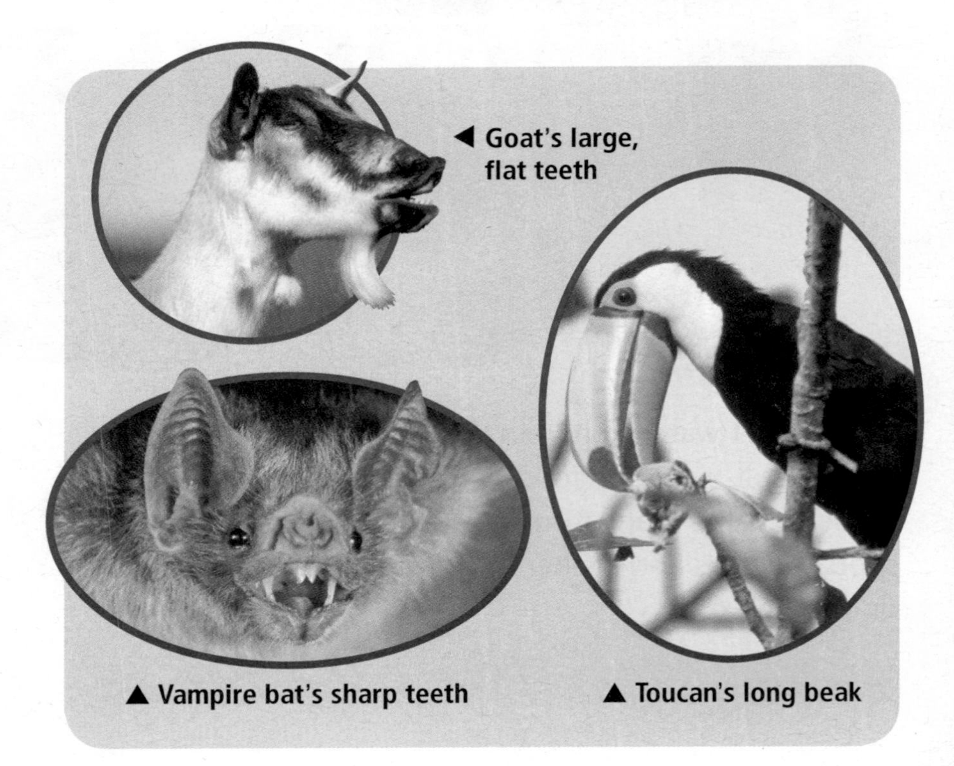

◀ Goat's large, flat teeth

▲ Vampire bat's sharp teeth

▲ Toucan's long beak

The snowshoe hare's fur color is an example of camouflage. In summer, its fur is rusty brown, like the rocks and ground. In winter, its fur is white, like the snow. These colors help the hare hide from predators.

Plants also have adaptations. One kind of adaptation is the kind of roots plants have. For example, mangrove trees grow in swamps. These trees have *prop roots* that help hold them up straight in the muddy soil.

▲ **The sundew plant has sticky hairs. They help the plant trap insects.**

Lesson Review

Complete this main idea statement.

1. Animals and plants have ____________ that help them meet their basic needs. ____________ is an adaptation that helps an animal hide.

2. Fill in this graphic organizer.

Main Idea: Animals have adaptations to help meet their needs.
Detail: An adaptation to meet the need for food: ____________ ____________
Detail: An adaptation to meet the need for water: ____________
Detail: An adaptation to meet the need for shelter: ____________ ____________

Georgia Performance Standard

S4L2a Identify external features of organisms that allow them to survive or reproduce better than organisms that do not have these features (for example: camouflage, use of hibernation, protection, etc.).

Vocabulary Activity

In this lesson, you will learn about ways in which animals behave and how behaviors help them survive in their environment.

Write the vocabulary word next to its definition.

	the movement of animals from one place to another and back
	a behavior with which an animal begins life
	when a kind of plant or animal has died out
	an inactive state

Student eBook
www.hspscience.com

Lesson 2

VOCABULARY
instinct
hibernation
migration
extinction

What Are Behavioral Adaptations?

An **instinct** is a behavior that an animal begins life with. Knowing how to build a nest is an instinct that birds have.

Hibernation is an inactive state. Hibernation helps this bat survive the winter.

Migration is the movement of animals from one place to another and back.

Extinction means that a kind of plant or animal has died out. Dinosaurs became extinct about 65 million years ago.

Insta-Lab

Observing Change

1. Look at old photographs of your community.
2. Look to see if any areas that used to be forest or other natural habitats have been built on.
3. In those areas that have been built on, how has it changed the way the area looks?

   ____

4. Think about the plants and animals that still live in the area. What kinds of adaptations do they have that allowed them to survive?

   ____

   ____

   ____

✓ Quick Check

Focus Skill The main idea of these two pages is Animals start life knowing how to do many things. Details tell more about the main idea. Underline two details about things that animals know how to do when they start life.

1. What are two animal instincts?

____

____

____

2. In the space below, draw an animal doing something from instinct.

Instincts

Animals are born knowing how to do many things. Spiders know how to spin webs. Zebras know that living in herds keeps them safe. Many birds know how to build nests. These behaviors are called instincts. An **instinct** is a behavior that an animal is born with. It helps the animal meet its needs.

◀ Spiders begin life knowing how to spin webs.

Hibernation

Many animals live where winters are cold. Often they know by instinct how to get ready for winter. They eat more. They find or build shelters. They hibernate there. During **hibernation**, an animal's normal body activity slows. Its heart barely beats. Its breathing almost stops. And its body temperature drops.

A hibernating animal uses little energy. It lives all winter on fat stored in its body.

▼ **Koi fish hibernate at the bottom of a pond.**

▼ **Woodchucks dig holes to hibernate in.**

✓ Quick Check

1. What happens to a hibernating animal's body?

___

2. Why don't hibernating animals have to eat?

___

___

3. Where do koi fish hibernate?

___

4. How do animals that hibernate get ready for winter?

___

✓ Quick Check

The main idea of these two pages is Migration is another instinctive behavior. Details tell more about the main idea. Underline two details about migration.

1. What is migration?

____

____

2. Name three animals that migrate.

____

3. Explain why some animals migrate.

____

4. Look at the graph on this page. Estimate how much farther the arctic tern migrates than the sandpiper migrates.

____

Migration

Migration is another instinctive behavior. **Migration** is the movement of animals in a group from one place to another and back.

Animals migrate for different reasons. They may move for more food. They may move to a better climate.

Caribou and whales migrate. They migrate each year at about the same times. These migrations depend on the seasons.

Gray whales migrate south in summer to give birth.

Which Animal Migrates the Farthest?

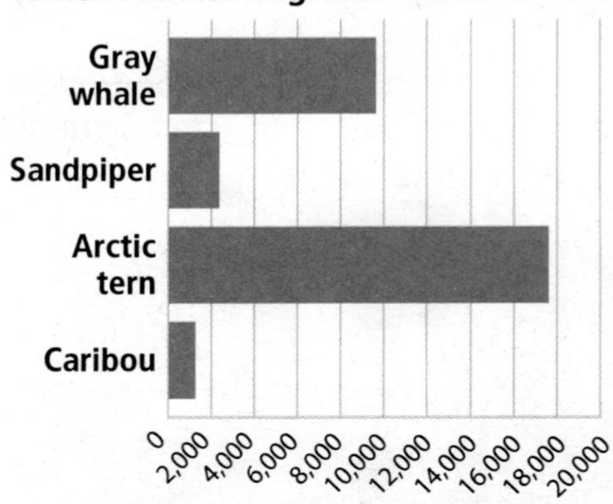

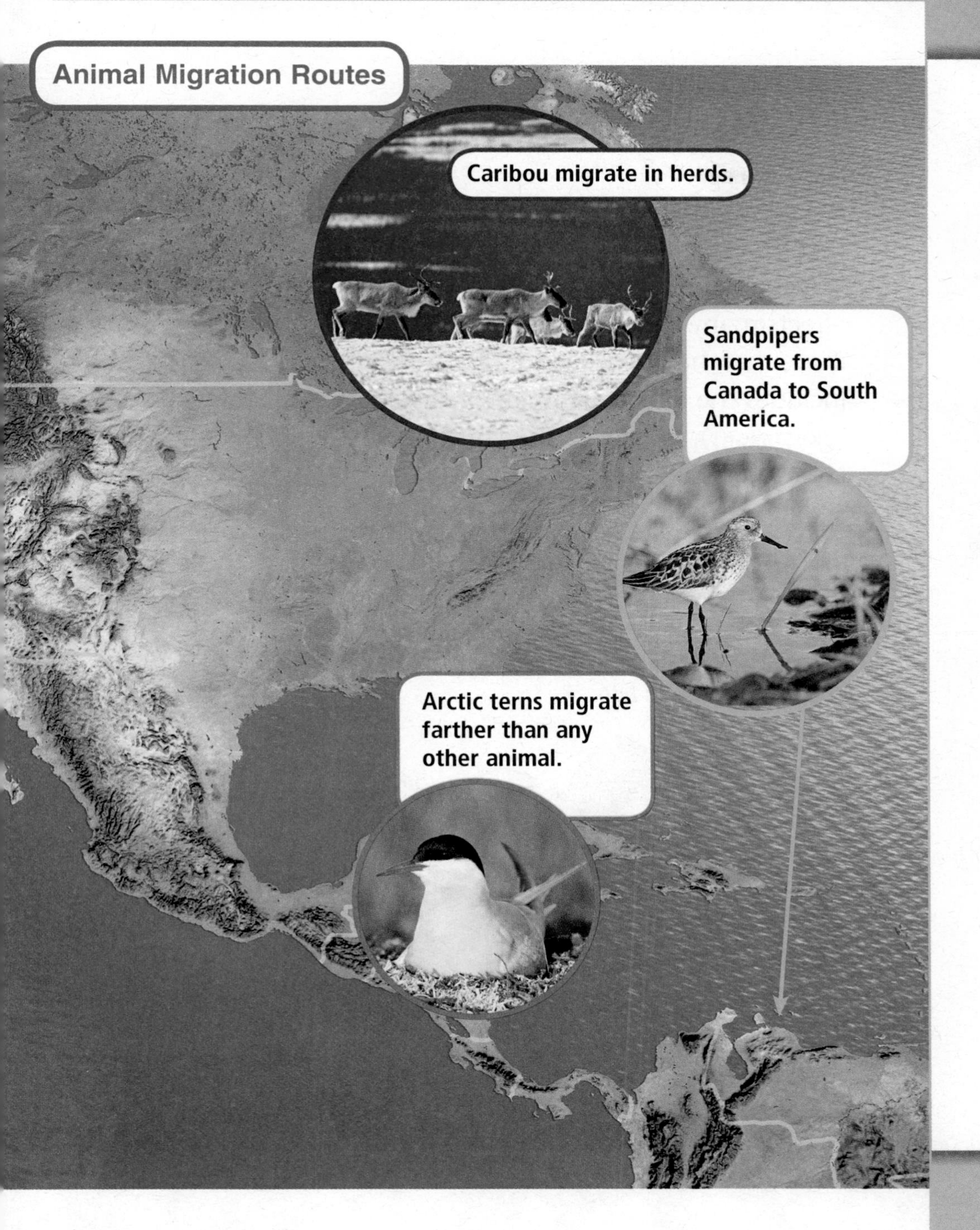

✓ Quick Check

Use the pictures on these pages to answer the following questions.

1. Place an **X** on the animal that migrates the farthest.
2. Why does the gray whale migrate?

   ____

3. When does the gray whale migrate?

   ____

4. What two locations does the sandpiper migrate from and to?

   ____

Quick Check

The main idea of these two pages is a change in the environment can be so great that it affects many living things. Details tell more about the main idea. Underline two details about how changes in the environment can affect living things.

1. Tell whether each statement about extinction is true or false.

   ______ Extinct animals are animals that can live only in one type of habitat.

   ______ People may cause extinction.

2. Why might a living thing become extinct?

____

____

3. What are two ways in which people can cause extinction?

____

____

Extinction

Plants and animals have adaptations that help them survive in their habitats. Suppose the habitat becomes drier or colder. Plants may die. Animals may also die if there is not enough food.

A change in the environment can be so great that it affects many living things. About 65 million years ago, more than half of all living things on Earth died out. This includes the dinosaurs. These living things became extinct (ek•STINGKT). **Extinction** means that all members of a certain kind of living thing have died.

The last woolly mammoth died about 30,000 years ago. ▼

Bachman's warbler was found throughout the Southeast. It was last seen in the early 1960s and is now thought to be extinct. ▼

The last saber-toothed cat died about 10,000 years ago.

How Humans Affect Ecosystems

People may cause extinction. People destroy habitats when they cut down forests or fill in wetlands. This can kill plants and cause animals to lose their homes. Animals may no longer have a place to get food.

Laws have been passed to help protect rare plants and animals. This is only one of the ways that people can help prevent extinction.

▲ **Hairy rattleweed lives only in Georgia, in the pine woods of Wayne and Brantley Counties. Because of logging, this plant is endangered.**

The gray bat is listed as endangered. It was known to live in northwest Georgia.

Lesson Review

Complete this main idea statement.

1. Animals begin life with ____________, or behaviors that help them survive.

2. Fill in this graphic organizer.

Main Ideas: Instinctive behaviors help animals meet their needs.

Everyday behaviors:
____
____
____
____
____

Migration:
____
____
____
____
____

Hibernation:
____
____
____
____
____

CRCT Practice

Fill in the circle in front of the letter of the best choice.

1. What is meant by the sentence *An organism is extinct*?

◯ A. All members of a certain kind of organism have died.

◯ B. Some members of a certain kind of organism have died.

◯ C. Some members of a certain kind of organism live in areas that are far apart.

◯ D. All members of a certain kind of organism live in the same area. S4L2b

2. Which of the following is an adaptation that allows animals to survive during cold weather?

◯ A. predation

◯ B. long legs

◯ C. hibernation

◯ D. webbed feet S4L2a

3. Which of the following behaviors is an instinct?

◯ A. a bear cub learning to climb a tree to avoid danger

◯ B. lion cubs imitating hunting behaviors of their mother

◯ C. birds migrating to the same location every year

◯ D. a dog responding to its owner's commands S4L2a

4. How does the beaver depend on the trees in this ecosystem?

◯ A. The beaver hides in the trees.

◯ B. The beaver uses tree branches to build a shelter.

◯ C. The beaver eats leaves and fruit from the tree.

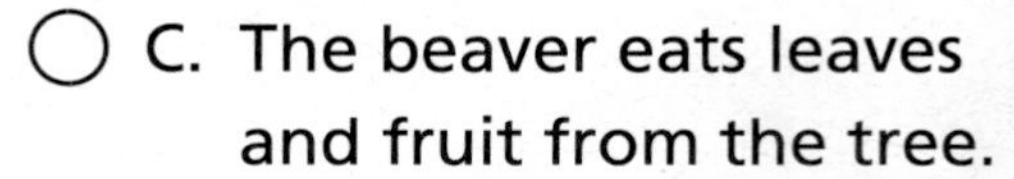

◯ D. The beaver gets honey from hives in the tree. S4L2a

5. Which of the following is NOT a cause of extinction?

- ◯ A. cutting down too many trees
- ◯ B. climate change
- ◯ C. passing laws to protect plants and animals
- ◯ D. filling in wetlands

S4L2b

6. A rare flower lives only in the woods behind a commercial area. If the woods are cut down so that businesses can be built, all of these rare plants will die. What would this be an example of?

- ◯ A. extinction
- ◯ B. migration
- ◯ C. adaptation
- ◯ D. competition

S4L2b

7. There are very few Bengal tigers left in the world. Bengal tigers are

- ◯ A. endangered.
- ◯ B. producers.
- ◯ C. extinct.
- ◯ D. herbivores.

S4L2b

8. Chipmunks have pouches in their cheeks. This allows them to store food and carry it back to their nests. What is this an example of?

- ◯ A. competition
- ◯ B. an adaptation
- ◯ C. migration
- ◯ D. an endangered species

S4L2a

9. Which of the following is a behavior that allows some animals to survive winter?

- ◯ A. competing
- ◯ B. swimming
- ◯ C. migrating
- ◯ D. crawling

S4L2a

10. Which is NOT an example of an adaptation being used?

- ◯ A. a rattlesnake disabling prey with venom
- ◯ B. a frog catching an insect with its long tongue
- ◯ C. a monkey choosing not to eat a plant that had made the monkey sick before
- ◯ D. a lion's sharp teeth cutting food

S4L2a

11. Which organism is extinct?

- ◯ A. Florida panther
- ◯ B. American eagle
- ◯ C. gray wolf
- ◯ D. woolly mammoth

S4L2b

12. The insect shown looks very much like a stick. This body shape is an example of

- ◯ A. camouflage.
- ◯ B. migration.
- ◯ C. extinction.
- ◯ D. hibernation.

S4L2a

13. Why don't you see any woodchucks in winter?

- ◯ A. They are migrating.
- ◯ B. They don't like the cold.
- ◯ C. They are hiding.
- ◯ D. They are hibernating.

S4L2a

14. There are very few eastern cougars left in the world. Eastern cougars are

- ◯ A. endangered.
- ◯ B. extinct.
- ◯ C. producers.
- ◯ D. herbivores.

S4L2b

15. When a predator frightens a skunk, the skunk sprays a foul-smelling liquid. This is an example of

- ◯ A. competition
- ◯ B. an adaptation.
- ◯ C. migration.
- ◯ D. extinction.

S4L2a

16. Which animal has the BEST camouflage for surviving life on a tree?

- ◯ A. a blue beetle
- ◯ B. a yellow bird
- ◯ C. a brown caterpillar
- ◯ D. an orange frog

S4L2a

17. A scientist observed these sea turtles. The scientist most likely said that their legs were adapted for

- ◯ A. swimming.
- ◯ B. catching fish.
- ◯ C. eating plants.
- ◯ D. walking on sand.

S4L2a

18. The arctic fox lives in the arctic regions of North America. In summer, its fur is reddish-brown. In winter, its fur turns white. How does this color change help the fox survive?

- ◯ A. White fur keeps the fox warm during the winter.
- ◯ B. The two colors of fur help the fox blend with its surroundings.
- ◯ C. White fur would absorb too much of the sun's energy during the summer.
- ◯ D. Reddish-brown fur would not keep the fox warm enough during the winter.

S4L2a

19. Which of the following is an adaptation that helps animals survive in hot, dry climates?

- ◯ A. hunting at night
- ◯ B. thick fur in bears
- ◯ C. a layer of fat on seals
- ◯ D. wings on birds

S4L2a

Use the drawing below to answer question 20.

20. Which is the BEST reason that an adult porcupine does not have many predators?

- ◯ A. It has sharp quills.
- ◯ B. It moves very slowly.
- ◯ C. It can swim in fast-moving water.
- ◯ D. It migrates when the weather turns cold.

S4L2a